REMAPPING SOUND STUDIES

# Remapping Sound Studies

GAVIN STEINGO AND JIM SYKES, EDITORS

.....

Duke University Press
*Durham and London*
2019

Text designed by Kim Bryant
Typeset in Merope by Westchester Publishing Services

Library of Congress Cataloging-in-Publication Data
Names: Steingo, Gavin, [date] editor. | Sykes, Jim, [date] editor.
Title: Remapping sound studies / Gavin Steingo and Jim Sykes, editors.
Description: Durham : Duke University Press, 2019. |
Includes bibliographical references and index.
Identifiers: LCCN 2018035544 (print)
LCCN 2018050317 (ebook)
ISBN 9781478002192 (ebook)
ISBN 9781478000372 (hardcover : alk. paper)
ISBN 9781478000464 (pbk. : alk. paper)
Subjects: LCSH: Sound (Philosophy) | Sound—Social aspects. | Sounds—Social aspects. |
Noise—Social aspects. | World music. | Music—Philosophy and aesthetics.
Classification: LCC B105.S59 (ebook) | LCC B105.S59 R46 2019 (print) | DDC 304.2—dc23
LC record available at https://lccn.loc.gov/2018035544

Cover art: Crater dunes and gullies, Mars. Source: NASA / JPL-Caltech / University
of Arizona.

# Contents

Acknowledgments, vii

Introduction:
Remapping Sound Studies in the Global South, 1
*Gavin Steingo and Jim Sykes*


PART I: THE TECHNOLOGY PROBLEMATIC
.....

1 ] Another Resonance: Africa and the Study of Sound, 39
*Gavin Steingo*

2 ] Ululation, 61
*Louise Meintjes*

3 ] How the Sea Is Sounded:
Remapping Indigenous Soundings in the Marshallese Diaspora, 77
*Jessica A. Schwartz*


PART II: MULTIPLE LIMINOLOGIES
.....

4 ] Antenatal Aurality in Pacific Afro-Colombian Midwifery, 109
*Jairo Moreno*

5 ] Loudness, Excess, Power:
A Political Liminology of a Global City of the South, 135
*Michael Birenbaum Quintero*

6 ] The Spoiled and the Salvaged: Modulations of Auditory Value
in Bangalore and Bangkok, 156
*Michele Friedner and Benjamin Tausig*

7 ] Remapping the Voice through Transgender-*Hījṛā* Performance, 173
*Jeff Roy*

PART III: THE POLITICS OF SOUND
.....

8 ] Banlieue Sounds, or, The Right to Exist, 185
*Hervé Tchumkam*

9 ] Sound Studies, Difference, and Global Concept History, 203
*Jim Sykes*

10 ] "Faking It":
Moans and Groans of Loving and Living in Govindpuri Slums, 228
*Tripta Chandola*

11 ] Disorienting Sounds:
A Sensory Ethnography of Syrian Dance Music, 241
*Shayna Silverstein*

12 ] Afterword: Sonic Cartographies, 261
*Ana María Ochoa Gautier*


Contributors, 275
Index, 277

## Acknowledgments

The project benefited greatly from an event held at Duke University in April 2016. We owe a huge debt of gratitude to Louise Meintjes for acquiring funding for, and largely organizing, the Remapping Sound Studies symposium. This would not have been possible without the generosity and organizational skills of three Duke University graduate students: Joella Bitter, Mary Caton Lingold, and Silvia Serrano. Thank you! We also thank the following institutions and groups within Duke University for their cosponsorship: the Franklin Humanities Institute (for a Humanities Futures Working Group award), the Department of Cultural Anthropology, the Concilium on Southern Africa, and the Ethnomusicology Working Group.

Joella Bitter, Rey Chow, Pedro Lasch, Mary Caton Lingold, Chérie Ndaliko, Marcelo Noah, and William Villalba offered invaluable feedback at the symposium through formal comments on papers. The contributors to this volume learned a lot from this thoughtful commentary. Many thanks to Jace Clayton, who delivered a stimulating lecture at the conclusion of the event.

At Duke University Press, Ken Wissoker did an incredible job shepherding this volume through to production. We thank him for his patience through many rounds of revisions and for always holding us to the highest standards. Elizabeth Ault was a tremendous force throughout the process. Thanks, Elizabeth, for your consistent support. Several anonymous reviewers also offered invaluable feedback along the way. We are very grateful to you, whoever you are. Special thanks go to Lee Veeraraghavan for producing the index and to Christopher Catanese for his stellar work as project editor. And thanks to Jonathan Sterne, who read and offered feedback in his typically generous and encyclopedic style.

At the University of Pennsylvania, several graduate students kindly read and offered comments on our introduction: Elizabeth Bynum, Elise Cavicchi, and Shelley Zhang. We also benefited from a graduate seminar at Princeton University.

Finally, we thank several colleagues and friends for support and stimulating conversations: Marié Abe, Kofi Agawu, Étienne Balibar, Georgina Born, Alexandrine Boudreault-Fournier, Manuel DeLanda, Kyle Devine, Bill Dietz, Steven Feld, Lauren Flood, Roger Mathew Grant, Jocelyne Guilbault,

Marilyn Ham, Wendy Heller, Helen Kim, Brian Larkin, Achille Mbembe, Rosalind Morris, Simon Morrison, Carol Muller, David Novak, Sarah Nuttall, Benjamin Piekut, Ronald Radano, Guthrie Ramsey, Timothy Rommen, Matt Sakakeeny, Martin Scherzinger, Elizabeth Schupsky, Gregory Smith, Will Straw, Jeffrey Snyder, Didier Sylvain, Peter Szendy, Gary Tomlinson, Daniel Trueman, and Barbara White.

*Acknowledgments*

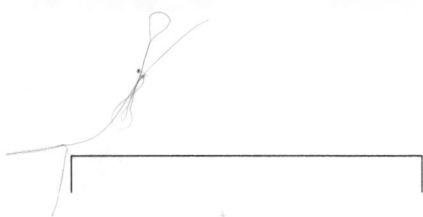

# Introduction

REMAPPING SOUND STUDIES

IN THE GLOBAL SOUTH

*Gavin Steingo and Jim Sykes*

. . . . .

## SOUND AND SOUTH

From at least the time of Jean-Jacques Rousseau, Western thought has associated sound with "South." Rousseau averred that the origin of language in warmer, southern climes was connected to music and the natural inflections of the voice. The frigid harshness of the North, he argued, allowed for no such melodiousness of speech: there, communication resulted in lifeless words. Sound, body, and presence on the one hand; arid speech—in close proximity to the "dead letter" of writing—on the other.[1]

To be sure, Rousseau's notion of sound is very different from our own, just as his Mediterranean South is not equivalent to the so-called global South of the twenty-first century. And yet one cannot help but notice an uncanny historical continuity: sound and South run like intertwined red threads through modernity, like a double-helix DNA constituting our underground makeup. Ever since Rousseau, the South has been associated with sound, music, body, presence, nature, and warmth. The North, by contrast, sees itself as dominated by writing and vision—by a cultural coldness born of the snowcapped peaks of the Alps.[2]

For Rousseau, as for us, sound was at once an empirical phenomenon and a concept burdened with tremendous political weight. The same, of course, can be said of "the South," a term that continues to designate a (loose and vague) geographical location while simultaneously harboring multiple ideological connotations with little empirical relation to geography and space. Sound and the South are, importantly, relational figures: they function only in relation to what they are not. Whether the relationship is dialectical, supplementary, or hybrid, sound and the South are the Others of the visual and the North. And like poles in any binary opposition, "sound" and "South" can easily be substituted for multiple "Other" terms, including "nature," "woman," "native," "Africa," "black," "queer," and "disabled."[3]

As negative figures of their respective binary relations, sound and the South historically have been positioned as resistant to analysis. For example, to this day phenomena associated with sound—such as timbre and music—are often deemed ineffable. The global South, for its part, is often derided as stubbornly failing to obey the (supposedly rational) logic of the state, that sine qua non of Western modernity. The North is often presumed to be the home of rationality and science; the South, of irrationality and magic. Precisely because they seem to evade the epistemic grip of Western reason, sound and the South are frequently offered as radical *alternatives* to the dilemmas of modernity. The problem here, as many have noted, is that celebrating "sound and the South" *against* "vision and the North" reaffirms the binary opposition on which all of the terms depend.[4]

In his magisterial *The Audible Past*, Jonathan Sterne (2003: 14) notices that, in the West, the "differences between hearing and seeing usually appear in the form of a list." Sterne calls this list "the audiovisual litany," which includes dictums such as "Hearing is spherical; visual is directional"; "Hearing immerses its subject; vision offers a perspective"; and "Sound comes to us, but vision travels to its objects." To this litany, one could easily add, "Sound is Southern; vision is Northern."

The most important thing to observe about the audiovisual litany, Sterne says, is that it is *ideologically loaded*. Despite being ostensibly a description of secular modernity, the litany's debt to Judeo-Christian theology is incontestable: the audiovisual litany is "essentially a restatement of the longstanding spirit/letter distinction in Christian spiritualism" (Sterne 2003: 16). As a deeply ideological and theologically inflected construct, the audiovisual litany is largely responsible for producing a calculus of value according to which sound or vision are—depending on your subject position—variously considered "good," "bad," "pure," or "impure."

But Sterne warns us against thinking of sound or vision as good or bad in and of themselves. If Western modernity is guilty of commodification and reification, this is not due to an overreliance on the eye and the gaze, as many have argued. "The primacy of vision cannot be held to account for the objectification of the world," writes Timothy Ingold. "Rather the reverse; it is through its co-option in the service of a peculiarly modern project of

*Steingo and Sykes*

objectification that vision has been reduced to a faculty of pure, disinterested reflection, whose role is merely to deliver up 'things' to a transcendent consciousness" (Ingold 2000: 235).

The Rousseauist equation of sound and the South haunts twentieth-century writing.[5] Much research on non-Western cultures, including research that claims to be scientific or empiricist in nature, reaffirms the ideology of the audiovisual litany. From Edmund Carpenter, Walter Ong, and R. Murray Schafer all the way (arguably) to Claude Lévi-Strauss and Marshall McLuhan, non-Western peoples are positioned as closer to sound and hearing than their European counterparts. It is no coincidence, then, that some of the twenty-first century's most prominent theorists of sound—Veit Erlmann, Jonathan Sterne, and Peter Szendy—begin with Jacques Derrida's famous deconstruction of binary oppositions (voice and writing, sound and vision, presence and distance).[6] The first move in any critical discourse on sound is to denaturalize and de-essentialize it.

Like "sound," "South" is a nebulous term that oscillates between an empirical category and ideological construct. The term "global South" is clearly not synonymous with the Southern Hemisphere, especially when one considers politically powerful settler colonies such as Australia, on the one hand, and, on the other hand, the existence of destitute ghettos located smack-dab in the middle of the world's richest countries. The term has also been criticized (sometimes even rejected outright) by people living in regions designated as the "global South," especially as a response to the term's deployment by international financial institutions such as the World Bank and the International Monetary Fund.[7] Others from "the global South" advocate a strategic appropriation of the term (see, e.g., Mahler 2015). All this being said, one cannot deny the coincidence (the "happening at the same time") of the European *idea* of the South as a place of poverty and naturalness and the expropriation and structural violence committed on large swaths of the Southern Hemisphere, including South America, Africa, South Asia, and Southeast Asia. The South is at once an idea, a socially constructed reality, and a partitioning of space. It is an ideology—a concept that appears in the form of an objective reality (Yuran 2014).

For the purposes of this book, we think of the global South as a set of global externalities produced through colonialism (of the "official" type, as well as contemporary settler colonies in North America and Palestine) and neoliberalism (particularly by imperialist practices such as "structural adjustment"). For reasons that are at once theoretical, heuristic, and strategic, we resist viewing any and all peripheries as part of the South. The vast majority of essays collected here address sound outside North America, Europe, and the advanced capitalist countries of northeastern Asia. The only exception is Hervé Tchumkam's chapter on the French *banlieues*, which we include because it shows precisely how contemporary France applies the logic of a Southern-trained colonialism within its borders.

If the South and sound are conjoined concepts lying at the heart of modernity, what might it mean to think these terms together in a way that does not reproduce colonial logic? *Remapping Sound Studies* proposes thinking sound not *as* the South (or as analogous *with* the South) but, rather, *in* and *from* the South.[8] Our approach consists of (1) an orientation toward ethnography and archives in diverse languages (including non-European languages) as ways to recognize that everyone—not only professional scholars—theorizes sound; and (2) a commitment to situating sound in and from the South not as a unified, alternative notion of what sound is but as diverse sonic ontologies, processes, and actions that cumulatively make up core components of the history of sound in global modernity.

What we are proposing is not simply a remapping of the dominant themes, narratives, and arguments of the heretofore Northern-focused field of sound studies *onto* the South; nor do we demarcate the South simply as a space for sonic difference. Rather, we develop a new cartography of global modernity for sound studies. This entails conceptualizing the South as a kind of radical horizon of geopolitics while dislodging the North as the site of the "original" and the "true" (Comaroff and Comaroff 2011). It also entails the "excavat[ion] of forgotten maps, imagining new ones or valorizing those that have been marginalized" (Vergès 2015). The invention of new cartographies requires, moreover, new ways to see and listen, as well as new navigational tools. From the birds and canoes of Marshallese "seascape epistemology" (chapter 3) and the natal charts used to generate Sinhala Buddhist (Sri Lankan) acts of sonic protection (chapter 9) to the oscillating tongues of Zulu ululators (chapter 2) and the fetal stethoscopes employed by Colombian midwives (chapter 4), *Remapping Sound Studies* offers myriad potent examples.

Are the narratives on the development of sound in global modernity that sound studies has told relevant for the global South? And are those narratives adequate as descriptions of modernity itself? Or do investigations into sound in the South allow us to challenge and reshape the conceptualization of modernity? These are some of the central questions raised in *Remapping Sound Studies* through case studies set in Southern locales.

We have deliberately framed this volume as an exercise in "remapping" rather than "decolonization" because we do not wish to overstate our contribution. Make no mistake: we wholeheartedly advocate all forms of decolonization, and we do draw on decolonization theory at strategic moments (more about this later). Our aims in this book, however, are rather more humble. What we hope to achieve is nothing more than laying out some potential alternatives to thinking about sound. We readily acknowledge this book's limitations: the contributors to *Remapping Sound Studies* are primarily based at Northern institutions (although not all of us are, and although some of us spend uneven amounts of time in different locales); there is also a general lopsidedness toward ethnography and anthropology (the result, in large part, of our extant professional network).[9] A more far-reaching project would require more money, more labor, and more infrastructure. It would include more writers based in the South; probably a few events in different geographical areas; and more writing and thinking in different languages, which would necessitate complex (and costly) mechanisms of translation. In short, what we have is not sufficient; one might even reasonably say that it is completely inadequate. We view this volume simply as an editorial effort that underscores the silence of the South in research on sound while beginning to map possible avenues for bringing Southern locales more clearly into sonic and auditory awareness.

The remainder of this introduction queries older and recent literature on sound in the global South to build a Southern-focused corpus for sound studies. We then articulate three domains—the technology problematic, sound's limits, and sonic histories of encounter—as basic grounds through which a Southern sound studies can fruitfully broaden (and, we argue, in some cases challenge) the narratives on sound in global modernity that have been produced to date by sound studies.

## WHITHER SOUND STUDIES?

In 2005, Michele Hilmes (2005: 249) wrote that the study of sound, "hailed as an 'emerging field' for the last hundred years, exhibits a strong tendency to remain that way, always emerging, never emerged." In some sense, the ten-plus years following Hilmes's observation have proved her wrong. Today, we can safely say that sound studies has fully emerged as a robust field, with its own journals, anthologies, and institutional positions. The sound studies boom has allowed for the useful recognition and claiming of disciplinary ancestors whose work had been recognized but had always appeared somewhat marginal to their respective disciplines (e.g., Attali 1977; Born 1995; Schafer 1977). Simultaneously, new vistas opened up for fields as disparate as the anthropology of the senses, the sociology of science, ethnomusicology, and aesthetics, leading to the further articulation of subfields (such as "ecomusicology") and the creation of research projects that could not have been glimpsed a decade earlier (see, e.g., Friedner and Helmreich 2012).

But there is another sense in which Hilmes is probably correct: disciplinary boom notwithstanding, the deep ideological framing of sound as the Other of vision will likely mean that it remains on the peripheries of knowledge. In terms of global hegemony, sound and South remain "marginal" almost by definition.

Noticing again the parallelism between sound and South, we are struck by the fact that the establishment of sound studies as a fledgling discipline has largely elided the global South. This lacuna is partially attributable to the fact that the sound studies boom has come largely from those working on the historical development of sound reproduction technologies; thus, emphasis has been placed on histories of technological innovation and progress.[10] This emphasis is closely associated with a certain homogenization of the listening subject (in much canonic work, he or she is white and middle class) and a tendency to flatten the sonic architecture of urban spaces, rendered simply as "global cities" or "the city."[11]

The neglect of Africa and Asia in recent anthologies and readers is striking. For instance, Routledge's four-volume *Sound Studies* anthology—containing seventy-two chapters and more than 1,500 pages (Bull 2013)—does not contain a single chapter on Africa or Asia (which together form more than half of the world's landmass and currently comprise more than one hundred sovereign nation-states).[12] Neither have any of the specific subgenres of sound studies that have emerged routinely focused on the global South, such as studies of sound in film (e.g., Coates 2008; Hilmes 2008); sound art and soundscape compositions (e.g., Cox and Warner 2004; Kelly 2011); or the recent wave of philosophical writing on "sound," "listening," and "noise" (e.g., Szendy 2008; van Maas 2015).[13]

One might assume that this neglect is attributable merely to the paucity of existing literature—that if we bracket work on *sound-related* topics (e.g., music, language), there is not very much rigorous research explicitly about sound in the global South. But even a cursory scan of the literature shows that this is not true. A bibliography for sound in the global South exists; it simply has not been integrated into the sound studies canon, save for a small number of well-known works.[14] It quickly becomes obvious, though, that once one sets out to incorporate a broader range of texts on the South

*Steingo and Sykes*

into sound studies, one finds *so many* texts that could reasonably be included that one's bibliography is quickly overwhelmed—so much so that it begs the question of what should constitute sound studies in the first place, much less a Southern-oriented version of it. Several bodies of literature, then, may not exactly "count as" sound studies, but they are certainly things to be aware of in any remotely comprehensive study of sound: work on sound in religious ritual (e.g., trance and funerary practices), the voluminous scholarship on music in India and Africa, and anthropological studies of speech acts related to the production of gender, power, and political oratory.

There is little to gain from employing the term "sound studies" for any and all literature that is even vaguely associated with sound. For example, we see little use in claims to the effect that musicology (or ethnomusicology) has "been doing sound studies all along."[15] Indeed, the histories of sound and music are relatively distinct: the philosophy of music has often deliberately elided sound, while scientific and medical approaches to sound historically have had minimal connection with music (see Barrett 2016). Taking a cue from Sterne, we therefore see much value in a *conjunctural* approach to sound that thinks various domains—musical, scientific, linguistic, theological, political—in relation to one another.[16] We are interested in an investigation that "gesture[s] toward more fundamental and synthetic theoretical, cultural, and historical questions" about sound (Sterne 2003: 5).

However, we hope to contest what exactly constitutes the so-called fundamental and synthetic questions of sound. Sound studies, as a disciplinary configuration, has begun to calcify around a specific—and, we would argue, quite narrow—set of concerns: the historical development in the West of "sound" as a concept and phenomenon separable from the other senses; the broader process of secularization and increased isolation of sound that was afforded by the invention of new technologies, including sound reproduction devices, medicinal technologies (such as the stethoscope), modern architectural spaces, and the science of acoustics; and the increasingly sharp division between public and private space. It is fair to say that these emphases produced an overarching *narrative* about sound in global modernity. We contend that a Southern sound studies will need to consider what the South "says" about this narrative; but we also suggest that to focus *only* on it would unwittingly turn sound studies into a story of Western influence on the South through the importation of the audiovisual litany and Western audio technologies—a neocolonial or imperialist narrative in which the West remains the protagonist.

As a way out of this deadlock, we propose listening to and from the South. This will bring to the fore a new set of conjunctures and raise a new set of questions. But listening from the South will also require insisting on the importance of several studies of sound that, for various definitional or ideological reasons, have been omitted from the quickly calcifying sound studies canon. We thus propose, both in this introduction and in the chapters that follow, using this literature as a jumping-off point for thinking sound and South together.[17] By rethinking the definitions of technology, of politics—and, indeed, of sound itself—*Remapping Sound Studies* listens backward and forward at the same time.

What follows is an "imaginary reader" that groups together Southern-focused sound studies literature that has been largely omitted from the sound studies canon. Perhaps the anthology will someday be published; perhaps not. As a thought experiment, however, our proposed "Southern Sound Studies Reader" serves two important functions: first, it highlights that substantive discourses on sound in the global South have long existed; and second, it begins to ask general (if not fundamental and synthetic) questions by demarcating key topics.[18]

The hypothetical reader consists of texts that had already been published when the extant anthologies were being compiled.[19] It therefore includes only work published before 2010.

---

*An Imaginary Southern Sound Studies Reader (First Attempt)*

. . . . .

### SOUND ECOLOGIES

Agawu, Kofi. "Rhythms of Society." In *African Rhythm: A Northern Ewe Perspective*, 8–30. Cambridge: Cambridge University Press, 1995.

Argenti-Pillen, Alex. "'We Can Tell Anything to the Milk Tree': Udahenagama Soundscapes." In *Masking Terror: How Women Contain Violence in Southern Sri Lanka*. Philadelphia: University of Pennsylvania Press, 2002.

Colombijn, Freek. "Toooot! Vroooom! The Urban Soundscape in Indonesia." *Sojourn* 22, no. 2 (2007): 255–73.[20]

Gupta, R. P. "Sounds and Street-Cries of Calcutta." *India International Centre Quarterly* 17, nos. 3–4 (1990–91): 209–19.

Mbembe, Achille. "Variations on the Beautiful in Congolese Worlds of Sound." *Politique Africaine* 100 (2005–2006): 71–91.

Morris, Rosalind. "The Miner's Ear." *Transition* 98 (2008): 96–115.

Ochoa Gautier, Ana María. "García Márquez, Macondismo, and the Soundscapes of Vallenato." *Popular Music* 24, no. 2 (2005): 207–22.

## SPEECH ACTS AND ORATORY

Bate, Bernard. "The King's Red Tongue." In *Tamil Oratory and the Dravidian Aesthetic: Democratic Practice in South India.* New York: Columbia University Press, 2009.
Brathwaite, Kamau. "History of the Voice." In *Roots.* Ann Arbor: University of Michigan Press, 1993.
Shelemay, Kay Kaufman. "Notation and Oral Tradition." In *The Garland Handbook of African Music,* vol. 1, ed. Ruth M. Stone. New York: Garland, 2000.
Tambiah, Stanley. "The Magical Power of Words." *Man* 3, no. 2 (1968): 175–208.

## RACE, ETHNICITY, CLASS, AND GENDER

Chandola, Tripta. "The Mingling of Modalities of Mingling: Sensorial Practices in the Karimnagar Slums." In "Listening into Others: In-between Noise and Silence." Ph.D. diss., Queensland University of Technology, 2010.
Marsden, Magnus. "All-Male Sonic Gatherings, Islamic Reform, and Masculinity in Northern Pakistan." *American Ethnologist* 34, no. 3 (2007): 473–90.
Minh-ha, Trinh T. "Mother's Talk." In *The Politics of M(O)thering,* ed. Obioma Nnaemeka. London: Routledge, 1997.
Weidman, Amanda. "Gender and the Politics of the Voice: Colonial Modernity and Classical Music in South India." *Cultural Anthropology* 18, no. 2 (2003): 194–232.

## SONIC ONTOLOGIES AND RELIGIONS

Becker, Judith. "Time and Tune in Java." In *The Imagination of Reality: Essays in Southeast Asian Coherence Systems,* ed. Alton L. Becker and Aram A. Yengoyan. Norwood, NJ: Ablex, 1979.
Feld, Steven. "The Boy Who Became a Muni Bird" (1979). In *Sound and Sentiment: Birds, Weeping, Poetics, and Song in Kaluli Expression.* Durham, NC: Duke University Press, 2012.
Hirschkind, Charles. "The Ethics of Listening: Cassette-Sermon Audition in Contemporary Egypt." *American Ethnologist* 28, no. 3 (2001): 623–49.
Nzewi, Meki. "African Musical Arts Creativity and Performance: The Science of Sound." *Nigerian Music Review* 6 (2005): 1–8.
Stoller, Paul. "Sound in Songhay Possession." In *The Taste of Ethnographic Things: The Senses in Anthropology.* Philadelphia: University of Pennsylvania Press, 1989.

## COLONIALISM AND NEO-COLONIALISM:
### ENCOUNTERS AND DOMINATION

Achebe, Chinua. "An Image of Africa: Racism in Conrad's *Heart of Darkness*." *Massachusetts Review* 18, no. 4 (1977): 782–94.

Roberts, Michael. "Noise as Cultural Struggle: Tom-Tom Beating, the British and Communal Disturbances in Sri Lanka, 1880s–1930s." In *Mirrors of Violence: Communities, Riots and Survivors in South Asia*, ed. Veena Das. Delhi: Oxford University Press, 1990.

DiPaolo Loren, Diana. "Beyond the Visual: Considering the Archaeology of Colonial Sounds." *International Journal of Historical Archaeology* 12, no. 4 (2008): 360–69.

Kun, Josh D. "The Aural Border." *Theatre Journal* 52, no. 1 (2000): 1–21.[21]

Tomlinson, Gary. "Unlearning the Aztec *Cantares* (Preliminaries to a Postcolonial History)." In *Object and Subject in Renaissance Culture*, ed. Margreta de Grazia, Maureen Quilligan, and Peter Stallybrass. Cambridge: Cambridge University Press, 1996.

### TECHNOLOGY AND MEDIA

Baucom, Ian. "Frantz Fanon's Radio: Solidarity, Diaspora, and the Tactics of Listening." *Contemporary Literature* 42, no. 1 (2001): 15–49.[22]

Larkin, Brian. "Degraded Images, Distorted Sounds: Nigerian Video and the Infrastructure of Piracy." *Public Culture* 16, no. 2 (2004): 289–314.

Lee, Tong Soon. "Technology and the Production of Islamic Space: The Call to Prayer in Singapore." *Ethnomusicology* 43, no. 1 (1999): 86–100.

Moisa, Lebona, Charles Riddle, and Jim Zaffiro. "From Revolutionary to Regime Radio: Three Decades of Nationalist Broadcasting in Southern Africa." *African Media Review* 8, no. 1 (1994): 1–24.

Pietz, William. "The Phonograph in Africa." In *Post-Structuralism and the Question of History*, ed. Derek Attridge, Geoffrey Bennington, and Robert Young. Cambridge: Cambridge University Press, 1987.

---

In addition, a "Southern Sound Studies Reader" would have to include new translations of texts from other languages. From Spanish, for example, the reader might include texts by Jesús Martín-Barbero, Orián Jiménez, Ángel Rama, and Julio Ramos.[23]

Important, explicitly Southern-focused interventions since 2010 include work by Lucas Bessire and Daniel Fisher (2012), Andrew Eisenberg (2013),

Ana María Ochoa Gautier (2014), Laura Kunreuther (2014), and Alejandra Bronfman (2016).[24] The edited volume *Keywords in Sound* (Novak and Sakakeeny 2015) is a powerful attempt at a truly global synthesis of sound theory—although this volume, too, is paradigmatically geared toward Euro-America.[25] Another important recent volume is *Audible Empire* (Radano and Olaniyan 2016), which provides a much needed consideration of how music was and remains integral to the circulation of "imperial logics" around the world. We see *Remapping Sound Studies* as participating in the same intellectual moment as *Audible Empire*, but our emphasis is more overtly on the global South, and we are less invested in "music" as a fundamental category of analysis. Indeed, it is possible to conceptualize these three volumes— *Keywords in Sound*, *Audible Empire*, and *Remapping Sound Studies*—as a kind of trilogy that carves out a crucial space in twenty-first century thinking about sound.

## SOUND STUDIES REMAPPED

Considering the most serious limitations of "sound studies" as a rapidly calcifying discipline, as well as the six key areas that emerge from our imaginary reader, it is possible to begin constructing a new cartography of sound theory. While there is, of course, no single way to remap sound studies, the chapters in this volume engage a coherent set of concerns. *Remapping Sound Studies* makes three principal proposals:

1. *Sound's relationship to technology.* We propose a shift from a focus on technology as a "modern" Western practice that reproduces, isolates, and idealizes sound to an analysis of "constitutive technicity" (Gallope 2011)—that is, of any and every supplement that humans engage in the production, reception, transduction, and attenuation of sound. In other words, we argue for a shift from technology as a set of inventions developed at a particular place and historical juncture to an exploration of the infinite series of objects and techniques through which "culture" is always already constituted.

2. *The question of sound as a relationship between listener and something listened to.* Sound necessitates a listener but *also* something *heard*. To say that something is heard means that there is some "thing" beyond and preceding human perception. In other words, the issue is not only a sensory one. It is also resolutely ontological, because the various peoples of the world understand that which is heard in radically different

manners. Thus, we propose viewing sound studies as an experiment with the thresholds or limits of audibility rather than simply a consideration of sound as a historically contingent "social construction." What we have in mind is a perspective that at once acknowledges the ontology of sound from a noncorrelational perspective (i.e., there exists an independent entity beyond human experience) *and* cultural differences in prehending sound. We suggest that ethnographies of the interrelations between these domains will form a critical component of a remapped sound studies.

3. *A conceptualization of sonic history as nonlinear and saturated with friction.* We propose that sonic history should be conceived as a narrative of jagged histories of encounter, including friction, antagonism, surveillance, mitigation, navigation, negotiation, and nonlinear feedbacks, rather than as efficiency, inexhaustibility, increasing isolation of the listening subject, and increasing circulation. Thus, in this volume we have incorporated a consideration of sound and the body—not only gendered sounds, but also how sound is used to listen in and through others and form social relations. This part of our project allows for politicized, historically situated, and culturally diverse narratives of sonic encounters in global modernity among variously defined peoples and their notions of sound.

Taken together, these three proposals suggest that sound studies can actively participate in remapping—if not exactly, or not yet, in decolonization proper—as an affirmative gesture and not simply as critique. We now turn to elaborating these three proposals, each of which culminates with a summary of how essays in this volume promote the proposals to work toward a comprehensive remapping of the study of sound.

### The Technology Problematic:
### A Proposal for Constitutive Technicity

Much initial work in sound studies as an intellectual field of inquiry was propelled by scholars working within science and technology studies and related disciplines such as communications and media theory (e.g., Bull 2007; Sterne 2003; Thompson 2002).[26] The focus of that work was the historical development of sound reproduction technologies, positioned as roughly analogous with Western "modern" devices. On the few occasions that "underdeveloped" areas of the world appear in the subsequent sound studies readers (e.g., Sterne 2012b), they are positioned mainly as laboratories for

*Steingo and Sykes*

exploring how Northern technologies spread.[27] Some readers are quite bold in their reduction of "sound studies" to "sound technologies"—for example, the *Oxford Handbook of Sound Studies* (Pinch and Bijsterveld 2011), which focuses exclusively on technology-related topics such as machine sound, tuning forks, ear plugs, sound in Pixar films, radio, and iPods.[28]

Because of the centrality of technology to canonical writings in sound studies and sound studies readers, many uses and encounters with sound by people in the global South that might otherwise count as sound studies are categorized differently, as *anthropology*—that is, as the study of Man.[29] But such a division is possible only if we assume an extremely limited view of both technology and anthropology. Taking a cue from Bernard Stiegler, we posit, first, that the term "technology" should refer to any technical object (not just so-called modern Western ones); and second, that there is no human prior to a relation with technical objects. In other words, there is no meaningful distinction between premodern, nontechnological humans and modern, technological ones. As Stiegler (1998: 193) remarks, "The being of humankind is to be outside itself." Humans are always already constituted through relations to technical prostheses. Accepting this insight would imply a realignment of sound studies' boundaries, since in our view the eighth-century Mexica song volutes described by Gary Tomlinson (2007), for example, are resolutely technological. Furthermore, by following the important move to rethink the distinction between nature and culture, the given and the made (Latour 1993; Ochoa Gautier 2006, 2014; Viveiros de Castro 1998), it is possible to think even of waterfalls—which, as Steven Feld (1996) explains, are constitutive of Kaluli acoustemology—as technical prostheses, thus further opening up possible terrain for sound studies.

The kind of conceptual tendencies we have noted are symptomatic of "a group of scholars operating within a field of discourse, an intellectual space defined by Euro-American traditions of ordering knowledge" (Agawu 2003: 58). Milisuthando Bongela defines the global South as a geopolitical space in which privileged bodies historically have constituted themselves through "naturalist study" (see Motlatsi 2016). We reject this naturalist epistemology and argue strongly for alternatives. Feld (2015) echoes these concerns in an impassioned statement. "What is most problematic to me about 'sound studies'," he writes, "is that ninety-five percent of it is sound technology studies, and ninety-five percent of that is Western. So if I refuse 'sound studies' it is because I think plants, animals, and humans everywhere are equally important to technologies, and I think that studying dynamic interactions of species and materials in all places and times are equally important and should be equally valued" (1).

Heeding Feld's warning, we are invested in a double maneuver. First, we are certainly in favor of studies that take seriously the dense interplay between humans and their multiple constitutive "outsides" (Ochoa Gautier 2014), whether through multispecies ethnography (Kirsky 2014), critical debates surrounding divine potency (Graeber 2015; Viveiros de Castro 2015), or various other possible means. But the turn toward (nonhuman) animal sound and various other sonic "ecologies" should in no way encourage a lapse into nativism. For this reason, we suggest looking at how "global" technologies are localized: regional social media platforms (such as South Africa's Mxit) or the specific entanglement of WhatsApp and Hindu nationalism in contemporary India, to provide just two examples.[30]

One aim, then, is simply to broaden the scope of what counts as a "sound technology." But we also contend that studies of "advanced" media and technology harbor problematic *theoretical* assumptions. For example, researchers tend to assume that, because of technological advances, music is becoming increasingly ubiquitous, moving at an ever faster pace in an unimpeded flow (Gopinath and Stanyek 2014; Kassabian 2013). Consider, for instance, that the aim of the *Oxford Handbook of Mobile Music Studies* (Gopinath and Stanyek 2014) is to examine "how electrical technologies and their corresponding economies of scale have rendered music and sound increasingly mobile-portable, fungible, and ubiquitous."[31] But why assume in the first place that technologies and their corresponding economies have rendered music and sound increasingly mobile-portable, fungible, and ubiquitous? By stating from the outset that authors for the *Handbook* examine *how* music has become more mobile, for example, the editors foreclose the possibility that in some places music has *not* become more mobile, or even, perhaps, that it has become *less* mobile. While the assumption about music's increasing mobility, fungibility, and ubiquity may be true of the global North (although even this is contestable), there are many contexts in Africa, Latin America, and Asia where technologies are marked by interruption, obduracy, and failure. *Remapping Sound Studies* does not aim to show *how* any given X (e.g., technology) has rendered music and sound Y (e.g., more mobile). Instead, it aims to radically expand the X so we can reorient the Y, so to speak.

Granted, sound and media studies researchers have paid increasing attention to technological failure in recent years, but the inclination is to look at failed devices to enrich or undermine dominant narratives about the technologies that *did* work (thus the emphasis on "quirky" objects). But as Brian

Larkin (2004: 291) has argued, "The inability of technologies to perform the operations they were assigned must be subject to the same critical scrutiny as their achievements." For Larkin, technological "inability" does not refer simply to quirks in a narrative structured around "modernity." Instead, his point is that failure and imperfection have *generative* as well as negative effects and that these effects are important in and of themselves (see also Morris 2010; Steingo 2015, 2016). This insight is particularly applicable in certain parts of the world—such as Kano, Nigeria, where Larkin conducted fieldwork—where technological failure or imperfection is a quotidian and normal part of life. Through the constant reduplication of analogue video-cassettes in Kano, sound is distorted to the point of unintelligibility. The people in Larkin's account should by no means be conflated with middle-class Northern consumers who deliberately engage degraded, broken, out-moded, or remediated devices (Bijsterveld and van Dijck 2009), since the ho-rizon of perfectly functioning technology is completely different in the two cases. But the point is precisely to explore different horizons or aspirations, different conceptions of what sound is and what it should be.[32]

Our proposal is therefore to think sound in relation to a wide range of technical supplements that constitute, rather than simply enhance, culture and history. At the same time, *Remapping Sound Studies* proposes a reassess-ment of the tropes of sonic mobility and fungibility, offering historically emplaced studies that engage issues such as failure, friction, and excess. As Gallope (2011: 49) suggests, "risk and failure" are also "*constitutive* of what it means to live." In short, we find that many of the elements largely absent from sound studies are in fact constitutive of any thinking about sound. This includes thinking about the global South in general: one does not think first about the global North and then, to go further or to be more politically correct, think also about the global South. Instead, the global South is constitutive of the global North—our task is to find ways to understand this constitutive process.

In his contribution to this volume, Gavin Steingo directly engages sound studies' technology problematic. For Steingo, the fact that sound studies' conclusions are based on evidence from Europe and North America is not in itself entirely problematic. What troubles him is that the cultural specific-ity of these conclusions is rarely acknowledged. Instead, certain observations are generalized, sometimes even becoming axiomatic. His chapter exam-ines three common arguments made by sound studies researchers—(1) that sound technologies are increasingly isolating the listening subject into indi-vidual "bubbles" (e.g., in automobiles and through mobile listening devices);

(2) that musical circulation is continually accelerating due to technological innovation and various forms of deregulation; and (3) that listening is associated with biopolitical investment and efficiency, as articulated, for example, by scholars dealing with attempts to combat workers' hearing loss in European industrial settings—and then places these arguments in dialogue with Southern contexts, particularly with the townships of South Africa, where he has conducted extensive fieldwork over the past ten years. Based on rigorous ethnographic evidence, Steingo challenges each of these arguments. Against the notion of audition in mobile bubbles, for example, he shows that in South Africa cars are both social and sonically "open" (see also Livermon 2008). Against accelerating circulation, he points to technological marginalization in the townships and maps out emerging forms of nonlinear sonic movement. And finally, he shows that hearing loss in South Africa's gold mines is characterized less by biopolitical investment than by logics of superfluity and abandonment (see also Morris 2008). Beyond simply critiquing the assumptions, methods, and conclusions of sound studies scholars, Steingo presents a grounded ethnographic study of a radically different relationship to sound.

Also focusing on South Africa, Louise Meintjes's contribution offers a quite different approach to sonic technicity. Meintjes undertakes a study of ululation as a vocal technique and as a form of acoustic reverberation that amplifies a woman's voice. Her essay remaps sound studies by focusing on reverberation—the *ululululu* of the ululator—as acoustic and relational, as African (and Middle Eastern), and as a metaphor for dialogue returning amplified and inflected from the South.[33] In particular, that dialogue shifts the attention in sound studies from technology to the technicity of voice; genders sound studies, thereby filling out the multiplicity of sound studies narratives; and finds sympathetic vibrations with black studies, which is also curiously underplayed in sound studies as it is evolving.

A third essay in part I addresses sound and technology in the global South. Jessica Schwartz examines sound reproduction and acoustical inscription in the Marshall Islands, as well as in the Marshallese diaspora. Through a close examination of heterogeneous technologies—from canoes and birds to radio programs and the mail—Schwartz puts pressure on orthodox notions of sound transmission and circulation. Among other significant insights, she shows that for her Marshallese interlocutors, sound is registered not as a scattered dispersal radiating outward (and dissipating gradually as it moves from the sender) but, rather, as a kind of connective tissue, a sociosonic *accumulation*.

Taken together, the contributions by Steingo, Meintjes, and Schwartz form a kind of disjunctive synthesis that goes some way toward remapping sound studies' technology problematic. The focus on South Africa—unintended and unanticipated by the editors of this volume—results in an intriguing constellation of related ideas and concepts. Schwartz's chapter, meanwhile, bolsters Meintjes's expansion of the definition of sound technology far beyond its normative usage.

### Listening at Sound's Limits: An Ontological Proposal

What are the limits of sound? Does it make sense of speak about sound beyond the threshold of a listening ear? Perhaps sound is not limited to audition (i.e., a sensory modality) but also encompasses the propagation of sounds by vibrating bodies prior to the audition of human perceivers. But can such a distinction be maintained? Can we move from sounds as they appear to us (i.e., phenomena) to sounds-in-themselves (i.e., noumena) without regressing into precritical naïveté? Perhaps twenty years ago such a move would have struck readers as hopeless, but in the wake of a renewed speculation (Bryant et al. 2011), the impulse to think sound outside of or beyond its human correlation is strong. Thus, for example, in a passionate and wide-ranging text, Steve Goodman (2009: 81) proposes an "ontology of vibrational force" that structures the entire cosmos and suggests that "sound is only a tiny slice, the vibrations audible to humans and animal." For Goodman, in other words, "sound" is what becomes of vibration when perceived by the (human and nonhuman) animal ear.

But how can we know what vibration is before it is transduced into sound by the ear? How can we know that it is, in fact, *vibration* if we know it only in relation to *us* as sound? Without succumbing to the viciousness of the "correlationist circle" (Meillassoux 2008), according to which "if I consider x, then *I* consider x always and only in relation to myself," there is no reason to assume with Goodman that what we call "sound" is really "vibration" prior to audition.[34] After all, anthropology is littered with examples of people hearing spirits and of shamanic travels to distant sonic worlds.

For our purposes, it is therefore enough to notice that sound is not identical to audition precisely because sound theorists often make claims about what lies beyond hearing—that is, what is *being heard*. In other words, sound studies is not reducible to the human sensorium, which means that sound studies is not identical to, or simply a branch of, sensory studies. What if we think of audition not in relation to the other senses but, rather, in terms of that which the auditory system intends or prehends? Given what Benjamin

Tausig (2013) terms the "posthumanist" emphasis of much sound studies,[35] it would be a mistake to fold sound studies into any cultural or ethnographic project that focuses exclusively on human perceptions, experiences, concepts, or sensations.

But if this is so, then where does that leave the remapping of sound studies? In our view, what the preceding discussion opens up is precisely the ontological stakes of sound.[36] But for us this does not mean simply positing vibration or some other figure as a unified ontological ground and then extrapolating humanly perceived sound as one minor hypostatization of that figure. It means, instead, taking seriously the existence of multiple ontologies, or as Eduardo Viveiros de Castro (2003: 18) famously put it, the "ontological self-determination of the world's peoples." Taking Viveiros de Castro seriously means recognizing multiple *natures* rather than multiple cultures — multinaturalism (perspectivism) rather than multiculturalism (relativism that presumes a "common nature or reality" [Ochoa Gautier 2014; Vanzolini and Cesarino 2014]).

While one task of remapping sound studies is to offer robust analyses of the different ways humans configure and relate to nonhuman sounds, another is to explore nonhuman sonic efficacy. Such a conjuncture of "natural and critical life" has much in common with the perspective mobilized by Anna Tsing (2015; see also Sykes 2018). As Elizabeth Povinelli (2016: 13) puts it, Tsing calls for "a more inclusive politics of well-being: a political imaginary which conceptualizes the good as a world in which humans and nonhumans alike thrive. And yet this thriving is, perhaps as it must be, measured according to specific human points of view, which becomes clear when various other species . . . come into view." Recognizing the difficulty and even the futility of such an inclusive politics, Povinelli (2016: 13) recommends that when exhaustion emerges from trying to solve the problem of universal inclusion, we focus instead on "local problems." This is a turn that several contributors make in this volume.

One way to capture local conjunctures between natural and critical life, without committing oneself wholeheartedly to some or other "ontological turn," is to pursue a rigorous analysis of variable thresholds or limits beyond which sound cannot be heard — what Jairo Moreno (2013: 215) calls a "general liminology." Rather than focusing on human audition or what lies beyond it, we advocate studies of the nexus through which audition is overwhelmed, exceeded, or repelled; elaborating on Moreno's work, then, we call for *multiple liminologies.*

*Steingo and Sykes*

Tripta Chandola (2010), for example, invites us to consider the permeability of the body in its relation to sound, pointing to the fundamentally relational character of auditory experience. Ana María Ochoa Gautier (2014: 8–9), taking a different approach, calls that which withdraws from audition the "spectrality of sound," asking us to consider the "excesses of the acoustic." But how precisely do we examine that excess? What is the excess exceeding? How is the limit constituted, and what, one might ask, is the limit of the limit? As a way to begin to answer such questions, we propose sound studies as an anthropological exploration of the multiply constituted limits of audibility.

In part II of this volume, chapters by Jairo Moreno, Michael Birenbaum Quintero, Michele Friedner and Benjamin Tausig, and Jeff Roy variously engage what we are calling multiple liminologies. Moreno brings his theoretical work on liminology to bear on an ethnographic case study of the listening practices of midwives in the Pacific Afro-Colombian region. He focuses on how midwives listen to the not-yet-born as "lives" that, because they are both human and near-divine according to local belief, have unique powers to listen in to the living. Listening, for Moreno's interlocutors, takes place at the threshold between the living and the nonliving—an ontological border zone he calls "quasi-life."

Like Moreno's, Quintero's essay in this volume focuses on Afro-Colombian populations and on the contingencies of the limits of sound. But in many ways, Quintero's context could not be more different: he explores the excessive, ear-splitting, bone-shaking volume of home sound systems as part of the musical aesthetic of the Pacific coast. Quintero begins by taking up theoretical arguments about excess and sovereignty from economic anthropologists and continental philosophers to suggest that loud music, as a site for the production of sensory excess, is part of people's (particularly men's) everyday performances of and micropolitical bids for prestige and sovereignty. In other words, *that which exceeds audition is constitutive of auditory experience.* Second, he examines a particular volume-related musical practice: raising volume to its technological limits while singing in groups at a throat-shredding full volume that is nonetheless inaudible below the sound pouring from the speakers. Quintero argues that in a local context of precariousness, violence, and seemingly perpetual impasse, musical volume functions as a kind of counterrepertoire to spoken language. This is necessitated by the fact that experiences of violence and endangerment in the Colombian Pacific are made banal by available registers of language (journalistic, confessional, political), even as life takes place in a temporality in which futurity is devoured by the permanent holding pattern of the precarious present. The unspeakability of violence and the inconceivability of a future without it belie psychological teleologies of trauma that identify memory, testimony, and witnessing as the ideal and even inevitable response to violence. In the breach between the unspeakability of violent experience and the inevitable incompleteness of attempts to repress it, practices of sonic excess provide a gestural, nonlinguistic, and nonliteral engagement with affect in an ambience marked by experiences of violence.

Next, Friedner and Tausig push the discussion of limits in a different direction in their collaborative study of deafness and value. Based on case studies from India and Thailand, they illustrate that sensory capacities are not biologically determined before a person steps into a network of cultural projects and local distinctions. Rather, capacity and value emerge within social, political, and economic contingencies. The border between hearing and deafness is therefore a variable one, and "disability" only ever marks the valuation of biological fact within a particular system of coordinates.

A final chapter in part II considers the entanglement of voice and sound with the gendering of bodies. Roy examines the crossing of sonic limits in a chapter on *hījṛā* performance, where the term "hījṛā" refers to transgender individuals in India. Roy suggests that the multifaceted queer and transgender-hījṛā (or trans-hījṛā) communities he works with invite us to understand the voice differently. He puts forth the claim that in trans-hījṛā contexts, the voice and its correlative identities should be understood outside the determinative framework of virtuosity and within the framework of *izzat* (roughly translated as "respect" vis-à-vis Gayatri Reddy [2005]), since it pertains to identity expressions that elude the stable logics of gender in which national and transnational identities are exchanged. Through case studies of vocal performance, Roy shows how trans-hījṛā communities sing—or otherwise "sound out" through uniquely stylized nonvirtuosic vocalic practices—as a means of generating respect among its members and transcending normative sonic spaces that engender normative behavior and identities. Situated explicitly within the volume's call for a turn toward the global South, Roy remaps the sonic understanding of identities that contest or ignore conventions of aural approval. The voices that Roy discusses do more than shift from one gender to another. They explode the very binary logic on which gender is constituted in the first place.

*The Politics of Sound:*
*A Proposal for Sonic Historiography as Encounter*

"Culture" is a mechanism of transduction, and it is a key hinge for relating to what is beyond human perceptibility. Studies of cultural constructions of sound in the global South can thus lay groundwork for thinking about sound generally and are indeed necessary for interpreting thresholds of audibility outside a Northern lens.

First, we need to question the veracity of the story of Western exceptionalism—that the West was the first to conceive of sound apart from the other senses.[37] Perhaps a study such as Erlmann's *Reason and Resonance* (which is a study of constructions of aurality and the ear in the Western philosophical tradition) is possible outside the West. What we are envisioning here is the possibility for textually based studies with a broad historical lens, something like Lewis Rowell's ([1992] 2015) work on South Asian musical manuscripts but looking not at discourses on "music" but at texts that locate non-Western traditions of thinking about "sound." Such a project will need to bear in mind the historical construction of "sound" as a category of Western modernity and not read that into non-Western materials. Rather, such materials should be read for their constructions of sound prior to and different from their engagements with Western notions.

Besides the articulation of sonic ontologies relating to distinct cultural zones and peoples, we propose that sound studies theorize sonic history as "encounters" constituted by friction and integration. One starting point is Andrew Sartori's (2008) argument that political liberalism and culturalism were two distinct but overlapping phases in early and late nineteenth-century India, respectively, which resulted in the conceptualization of "the arts" and "culture" apart from politics and economics. The latter were assigned to public space and gendered male, while the former were assigned to "private" domains such as the household and temples and gendered female (Birla 2009; Sartori 2008, 2014).[38] Placing sound within such colonial transformations will require attending to how the tendency for some sounds to become metonymic for certain kinds of people and traditions facilitated legal determinations of what sonic practices were deemed to belong in "public" and "private" spaces. Native sounds, as equivalent to South/female/domestic/indigenous (and so on), came to appear as disruptive intrusions into a public space that was defined as rightfully belonging to politics and the market (Sykes 2015, 2017).

The first chapter of part III, by Hervé Tchumkam, shows that such spatial divisions and the power of sound to act as a disruption can easily be mapped onto the relationship between today's ethnicized suburbs and the centers of global cities. Tchumkam explores the politicization of inaudibility in contemporary France, employing a study of sound to analyze the social, political, and historical conditions for urban riots. He begins by reflecting on the explosive events of 2005, when France was struck by violence in the *banlieues*, the projects on the outskirts of French cities that are populated mainly by African migrants and their offspring. Tchumkam's essay sheds light on the ways in which the banlieues become a primary site for the replication of colonial rules. At the same time, because they are contained within the state, these spaces represent a serious threat to the political order. Caught between inclusion and exclusion, the invisible and unheard citizens of France have turned into rioters for justice and equality. Ultimately, his study of the invisibility and *inaudibility* of "visible minorities" in a space relegated to the periphery of major French cities seeks to show that the only sounds the relegated periphery of France is left to produce is that of uprisings. By examining distributions of sensory perception — audible and inaudible, visible and invisible — Tchumkam focuses on the relationships between power, sound, and that which is not yet heard. While not technically about the global South, the paper addresses France as a postcolonial space that replicates the logic of coloniality within the metropolitan center. Tchumkam illustrates how an analysis of "postcolonial France" forces us to reconsider notions of justice, equality, voice, and sonic politics.

Jim Sykes's chapter explores the relationship among Theravada Buddhist sonic ontology, colonialism, and the history of Christian missionization. Sykes shows how the ritual practices of Sri Lanka's Sinhala Buddhist ethnic majority involve numerous forms of efficacious speech and sonic exchange with gods and demons, in conjunction with the placement of the stars, which function to protect and heal individuals and the population at large. To unlock these powers, sonic utterances (including drumming) must be made in certain directions and at certain times (determined by an astrologer), and may require the holding of specific objects. Sykes argues that Christian missionaries defined Buddhism as an ideally silent religion whose sonic elements consist mainly of the chanting of monks; thus, they found the religion's noisiness (as they witnessed it) to be the result of the religion's decline in the hands of Sinhalese and the infiltration of supposedly "Hindu" elements such as astrology and deity propitiation. Sykes warns that if scholars think they are doing sound studies simply by exploring sound in religious contexts in isolation from the other senses, objects, supernatural beings, astrology, and the like, they run the risk of reproducing a European ideology of "sound in itself" that, in the global South, has a specifically colonial and Christian heritage.

The final two chapters in part III focus explicitly on gender and embodiment. Tripta Chandola's essay explores the relationships among sound, womanhood, adulthood, memory, and social relations (male-female and female friendships) as they emerge for young women who live in three adjoining slum settlements (called Govindpuri) on the outskirts of the Indian city of Delhi. The anecdotes narrated by Chandola's interlocutors show how some young women in Govindpuri revel in manipulating men by using sound to "fake" certain emotions, such as moans and groans during staged phone sex in exchange for new iPhones. In another anecdote, young girls fool a female friend by facilitating a cassette mixtape exchange in which she thinks she is receiving a carefully constructed playlist of romantic songs from her imagined lover. In a third anecdote, Chandola and her female confidant stage a conflict with an older singing salesman in the local market, whom they accuse of lip syncing. Through Chandola's (and her consultants') recalling of incidents in which they "faked it" or accused others of "faking it" through sound, Chandola deftly shows the power of women to use sound for their own agency and how their successes and failures at "faking it" through sound were a key domain through which these Govindpuri girls played at being (and became) adults. Now married women with children of their own, many of Chandola's interlocutors now look back wistfully on those times.

Finally, Shayna Silverstein's chapter brings the discussion of the politics of sound to the level of the self, the body, the ethnographic encounter, and representation. In turn, she situates her memories of learning and experiencing Syrian *dabke*, a highly participatory and interactive popular type of dance music, in preconflict Syria and within a contemporary politics of the importance of expressive culture for Syria's refugee community. At its heart, Silverstein's essay rethinks the relationship between listener and sound object by engaging with nonauditory senses as crucial to the constitution of selfhood. Based on ethnographic fieldwork on performance dynamics in dabke, Silverstein stages several encounters between herself and her interlocutors that pivot on moments of sensory disorientation—that is, moments that reveal how the contingencies of lived experience entrain our bodies to perceive the world in culturally specific ways. In particular, she focuses on kinesthetics to raise questions about how proprioception, movement, and tactility direct bodies in sonically dense environments. Drawing on Sara Ahmed's queering of modes of perception, she suggests that ethnographic encounters disrupt the "habitual and elided" (Ahmed 2006) relations of bodies, space, and time. As a mode of performance ethnography, the work of sensory disorientation both engages in and disrupts interpersonal and intercorporeal acts of performance in ways that help us to better understand the cultural logics and performative processes that shape ethnographic subjectivity. Disorientation of the ethnographer's sensibilities thus informs how we approach sound and the insights we gain through its study because it accounts for body techniques such as listening and dancing as incidental, subject to spatiotemporal disjunctures, and indicative of social distinctions between researchers and their object of study. By embodying the ethnographic process through dabke practice, Silverstein deprivileges intellective modes of knowledge production and redistributes the senses in ways that challenge the disciplinary genealogy of sound studies.

## CLOSING REMARKS, OR, "THE SOUTH WAS THE PROMISE OF OTHER THINGS TO COME"

Remapping sound studies participates in a remapping—and, indeed, a partial decolonization—of thinking and listening. Drawing on Viveiros de Castro's (2004) notion of conceptual "equivocation," we advocate a "transformation or even disfiguration" (Holbraad et al. 2013) of thinking about sound, about ways of hearing, and about the constitution of entities that hear via Southern perspectives. To "remap" sound studies, then, means engaging potential equivocations head-on—listening across time and place in a manner that lives up to the challenges of twenty-first-century geopolitics.

"On a global scale," writes Françoise Vergès (2015), "following the mapping and remapping of what matters—and what does not—means following the routes of racial capitalism, the transformation of land into spaces for the working of capital" (28). But, she continues, "Historical and political cartographies mix with personal cartographies, building a multi-dimensional space of memories" (28). We quote Vergès's account of these intertwining cartographies at length:

Where and how I grew up gave me a cartography of global resistance to power, colonialism, and imperialism. From the Greek χάρτης, or "map," and γράφειν, "write," cartography is, of course, the art and science of drawing maps. My first geography of resistance was drawn by the Réunion Island anticolonial movement. It was from this small island in the Indian Ocean that I read the world. To the local cartography of cultural and political resistance, I added the millenary world of exchanges between Africa and Asia; the world of solidarity routes among anti-imperialist movements of the various Souths; the Southern world of music, literature, and images. (28)

Europe, Vergès writes, "was geographically and culturally on the periphery" (28). In her youth, "The South was the promise of other things to come . . . a map of third-world feminism, of national liberation movements, of the promise of Bandung" (28). But Vergès found this hopeful and potentially liberating "solid cartography" she inhabited during the 1960s and '70s crumbling as time went on (28). Yet even as the "mutilated and mutilating cartography" (40) created by racial capitalism persists, Vergès notes that we have entered a new era, "an environmental wasteland where media never die, and [have facilitated] a colonization of the self . . . in which new sites of forgetfulness are created, new Souths" (39). Conjuring up the hopefulness of her youth, she concludes, "I have still a South. I look for its emergence in the resistance to the constant process of territorialization and deterritorialization operated by racial capital" (33).

It is in this spirit that *Remapping Sound Studies* listens *for* and *from* the South, with the aim of resisting the unwitting convergence of sound studies' Northern-centric narrative and the ever mutilating cartographies of racial capital, by configuring sonic solidarities across Southern spaces defined by difference and agency.

### Notes

1. The key text here is Rousseau's *Essai sur l'origine des langues* (published in 1781). The secondary literature is vast. See, e.g., Starobinski 1989; Thomas 1995. On the relationship between culture and the environment in the history of Western thought, see Glacken 1976.

2. Similarly, nineteenth-century British writers described the Malay language as the "Italian of the East," referring to a sweet melodiousness in Malay that they found lacking in East Asian languages (Irving 2014).

3. For important work on this manifold construction, see Haraway 1991; Minh-ha 1989.

4. For a particularly potent and relevant critique, see Cimini 2011.

5. We focus here on twentieth-century social science, although much could be said about earlier periods and other modes of knowledge production: see, e.g., the recent, excellent critique of sound in nineteenth-century colonial literature in Napolin 2013, which echoes an earlier critique in Achebe 1977.

6. Deconstruction is treated very differently in the work of Sterne, Szendy, and Erlmann: see esp. Sterne 2003: 17–18; Szendy 2015: 18–19; Erlmann 2010: 14–16, 48–50.

For a relevant deconstructivist text, see Derrida [1967] 1976. For a useful critique of binary oppositions in Carpenter, McLuhan, and Ong from a different perspective, see Feld 1986.

7. For a critique along these lines, see the contributions to Roy and Crane 2005.

8. Important inspirations for this move include Comaroff and Comaroff 2011 (as well as the series of responses collected in Obbario 2012); Connell 2007; Santos 2014. For a useful set of reflections on cities in the global South, see the contributions in Dawson and Edwards 2004. The journal *Global South* (published biannually since 2007) is another important resource. For a different perspective, see Latimer and Szymczyk 2015.

9. Speculating on what types of theorization may yet emerge from other disciplinary perspectives is tantalizing indeed. For an excellent example of what a historian may do with Southern sound, see Bronfman 2016.

10. At the turn of the twenty-first century, important studies appeared on topics such as technological modernity (Sterne 2003, 2012b), architectural acoustics (Thompson 2002), and histories of hearing, listening, and aurality (Erlmann 2004, 2010; Szendy 2008) — to name just some of the more celebrated examples.

11. We recognize the importance of this scholarship (see, e.g., Bull 2007; LaBelle 2010). One of our aims is to put it in dialogue with writings on urban life, design, and spatiality outside the global North (see, e.g., Kusno 2010; Nuttall and Mbembe 2008; Simone 2009).

12. To provide a few other representative examples: *The Oxford Handbook of Sound Studies* (Pinch and Bijsterveld 2011) contains twenty-three chapters on topics ranging from Pixar and birdsong to cochlear implants and iPod culture, but Africa and Asia are absent there as well. The earlier edited collection *Hearing Cultures* (Erlmann 2004) is also North-centric, the lone exception being Charles Hirschkind's chapter on Egypt. Routledge's single-volume *Sound Studies Reader* (Sterne 2012b) fares slightly better: of its forty-five chapters, there is just one on southern Africa (by Louise Meintjes) and two on North Africa (one by Hirschkind and an early text by Frantz Fanon). David Novak and Matt Sakakeeny (2015: 7) make the point forcefully: "But despite the interdisciplinary breadth of sound studies, the field as a whole has remained deeply committed to Western intellectual lineages and histories. As one example, of the dozens of books about sound published by MIT Press — a leader in science and technology studies, philosophies of aesthetics, and cognition — none is principally invested in non-Western perspectives or subjects. Sound studies has often reinforced Western ideals of a normative subject, placed within a common context of hearing and listening."

13. This situation seems set to change, though, as recent years have witnessed a trickle of publications that theorize sound in specific locations of the global South (see, e.g., Bronfman and Wood 2012; see also Bronfman 2016). Another example is the recent issue on Southeast Asian soundscapes in *Journal of Sonic Studies*: see http://sonicstudies.org/JSS12.

14. These include the aforementioned works of Fanon, Meintjes, Hirschkind, and Ochoa Gautier (see note 12).

15. As ethnomusicologists, we have occasionally encountered this sentiment from colleagues in informal conversations.

16. Sterne (2012a: 3) writes, "To think sonically is to think conjuncturally about sound and culture. . . . Sound studies' challenge is to think across sounds, to consider sonic phenomena in relations to one another."

17. In doing so, we systematically elaborate a conjuncture alluded to in several recent texts (Novak and Sakakeeny 2015; Stadler 2010). Novak and Sakakeeny (2015) have also recently pointed to the Eurocentricism of music work on sound. We heed their call for increasing attention to a plurality of sonic practices.

18. We do not suggest our list is exhaustive; we have picked valuable texts that remain altogether unincorporated into sound studies. Each essay points toward topics with their own long bibliographies. Also, we have placed the voluminous literature on "music" off to one side but do include a few musicological sources that explicate broader sound-related topics under "sonic ontologies and religions." Finally, our reader should be placed in dialogue with a few subgenres of sound studies that can be seen as running parallel to this "hidden" Southern version of the discipline, such as recent writings on sound, listening and blackness in North America (see, e.g., Brooks 2010; Nyong'o 2014; Radano 2016; Stadler 2010; Stoever 2016; Weheliye 2000, 2014; White and White 2006) and the auditory turn in American studies (see, e.g., Daughtry 2014; Morat 2014; Schmidt 2002).

19. Extant anthologies include Bull 2013; Pinch and Bijsterveld 2011; Sterne 2012b.

20. See also Barbara Watson Andaya's (2011) essay on sound and power in the pre-modern Malay world, as well as the discussion on the Indonesian and Malay concept of *ramai* (busy noisiness) in Rasmussen 2010.

21. This piece is about aural constructions of the U.S.-Mexico border.

22. We include this text rather than Fanon's "This Is the Voice of Algeria" because the latter is included in some sound studies anthologies and because Baucom's piece theorizes the importance of radio for anticolonial movements broadly.

23. Also of interest is the Argentinian journal *El oído pensante*.

24. Also of note at the 2010 dividing line are two landmark articles in the *Annual Review of Anthropology* (Porcello et al. 2010; Samuels et al. 2010), both of which are relevant to (and have helped us conceptualize) the current introduction, but both of which ultimately have different aims to our own.

25. See the comments in this regard in Sterne 2015: 71, 74n3.

26. Note that we do not say science and technology studies was the *only* site for the crystallization of "sound studies" as a formal discipline in the early-to-mid-2000s. For example, parallel (and occasionally intersecting) lines of inquiry at that time include writings on sound and music in everyday life (see, e.g., Bull 2000; DeNora 2000) and sound art (see, e.g., Cox and Warner 2004; Kahn 2001; Kelly 2011; Kim-Cohen 2009).

27. We are thinking particularly here of why Fanon's famous text on radio in Algeria, Hirschkind's work on cassette sermons in Egypt, and Meintjes monograph on recording studios in South Africa are three of the only texts focusing on the South that are widely recognized as a part of sound studies.

28. It bears emphasizing that, while the scholarly study of sound art builds on a long history of experimenting with sound through technology, from the Italian futurists through John Cage, *musique concrète*, the development of the synthesizer, and so on, we suggest that neither these nor the scholarly writings about them should be reduced to or taken as equivalent to sound studies. Embedded in our critique of the technology problematic in this section, in other words, is a critique of the tendency for some writers to reduce sound studies to a narrative on experimenting with new musical instruments.

29. This may explain why Feld's work on the Kaluli is absent in the various sound studies readers.

30. On Mxit, see, e.g., Kreutzer 2009. On WhatsApp and nationalism in India, we have in mind Ravi Sundaram's keynote address the "What Is Comparative Media?" conference held at Columbia University in 2016. Another relevant example would be Sumanth Gopinath's (2013) study of cellphone ringtones, which is quite impressive in its geographical breadth.

31. This motivation is stated on the Oxford University Press website, accessed April 6, 2015, http://ukcatalogue.oup.com/product/9780195375725.do.

32. One study that seeks to mediate these different kinds of engagement with technological "failure" is "The Sublime Frequencies of New Old Media" (Novak 2011), on the importance of distortion to the aesthetics and processes of remediating "world music" in the digital age. Novak compellingly describes the investments in distortion made by Northern underground musicians since the 1980s — not just amplifiers and effects pedals but the circulation of cassettes — and the ways that distortion both is and is not valued in similar ways by those from Southern locations whose musics have been remediated for Northern markets in the digital age.

33. For a similar position, see Muller and Benjamin (2011) on musical echoes.

34. Brian Kane (2015) suggests that Goodman and a few other authors proclaiming the "ontological turn" in sound studies carry with them preconceived cultural notions about sound even as they proclaim to produce "culture-free analyses." See the following footnote for additional references to the contemporary debate.

35. "Posthumanist" may not, in fact, be the best or most precise word — at least in the way that we intend it. A more appropriate term for our own meaning would probably be "anti-correlationist."

36. A great deal of controversy surrounds the recent "ontological turn" in anthropology. Although often claiming a longer historical trajectory — going back at least to the 1960s and 1970s (e.g., Marilyn Strathern, Marshall Sahlins, Eduardo Viveiros de Castro) or, more controversially, to the structuralism of Claude Lévi-Strauss — so-called "ontological anthropology" has garnered several potent manifestos in recent

years (e.g., Holbraad and Pederson 2017), as well as numerous hard-hitting critiques (e.g., Bessire and Bond 2014; Graeber 2015). See Alberti et al. 2011; Venkatesan et al. 2010. See also the thread "The Politics of Ontology" on *Cultural Anthropology*'s Theorizing the Contemporary platform, https://culanth.org/fieldsights/461-the-politics-of -ontology. Anthropology's own ontological turn has taken place alongside a similar turn in continental philosophy, referred to variously as "speculative realism" (Bryant et al. 2011) and "new materialism" (Coole and Frost 2010). An invaluable contribution from the perspective of music and sound is Ochoa Gautier 2014.

37. For example, Adrien Tien (2015: 38) writes that the words *sheng* (sound), *yin* (sound), and *yue* (music) have been in use in China since the pre-Qin and Qin periods (before and up to 206 BCE). *Sheng* is "an acoustic stimulus generated by something in the environment, e.g., an event or an action, with no immediately identifiable agent" (Tien 2015: 49). For ancient Chinese philosophers, there was a distinction between sound and non-sound, and "both [are] equally valid aspects of sonic experience" (he remarks that "the word 'non-sound' is preferred over the word *silence* since . . . *silence* is an Anglo-centric word which does not have readily available, lexical and translational equivalents in other languages, including Chinese" [Tien 2015: 49]). Although this example is not from the global South, it points to the need for studies that consider how sound was configured in relation to the other senses in different locations around the world.

38. Amanda Weidman (2006), drawing on the earlier contributions of Partha Chatterjee, shows how "the female voice" became perceived as a site of an authentic and ancient Indian identity that was useful for the anticolonial movement, on account of this association of culture and women with the private sphere in India's colonial period.

## References

Achebe, Chinua. 1977. "An Image of Africa: Racism in Conrad's *Heart of Darkness*." *Massachusetts Review* 18, no. 4: 782–94.

Agawu, Kofi. 2003. *Representing African Music: Postcolonial Notes, Queries, Positions*. New York: Routledge.

Ahmed, Sara. 2006. *Queer Phenomenology*. Durham, NC: Duke University Press.

Alberti, Benjamin, Severin Fowles, Martin Holbraad, Yvonne Marshall, and Christopher Witmore. 2011. "'Worlds Otherwise': Archaeology, Anthropology, and Ontological Difference." *Current Anthropology* 52, no. 6: 896–912.

Andaya, Barbara Watson. 2011. "Distant Drums and Thunderous Cannon: Sounding Authority in Traditional Malay Society." *IJAPS* 7, no. 2: 19–35.

Attali, Jacques. 1977. *Bruits: Essai sur l'économie politique de la musique*. Paris: Presses universitaires de France.

Barrett, G. Douglas. 2016. *After Sound: Toward a Critical Music*. New York: Bloomsbury.

Bessire, Lucas, and David Bond. 2014. "Ontological Anthropology and the Deferral of Critique." *American Ethnologist* 41, no. 3: 440–56.

Bessire, Lucas, and Daniel Fisher. 2012. *Radio Fields: Anthropology and Wireless Sound in the 21st century*. New York: New York University Press.

Bijsterveld, Karin, and José van Dijck, eds. 2009. *Sound Souvenirs*. Amsterdam: Amsterdam University Press.

Birla, Ritu. 2009. *Stages of Capital: Law, Culture, and Market Governance in Late Colonial India*. Durham, NC: Duke University Press.

Born, Georgina. 1995. *Rationalizing Culture: IRCAM, Boulez, and the Institutionalization of the Musical Avant-Garde*. Berkeley: University of California Press.

Bronfman, Alejandra. 2016. *Isles of Noise: Sonic Media in the Caribbean*. Chapel Hill: University of North Carolina Press.

Bronfman, Alejandra, and Andrew Grant Wood, eds. 2012. *Media, Sound, and Culture in Latin America*. Pittsburgh: University of Pittsburgh Press.

Brooks, Daphne A. 2010. "'Sister, Can You Line It Out?': Zora Neale Hurston and the Sound of Angular Black Womanhood." *Amerikastudien/American Studies* 55, no. 4: 617–27.

Bryant, Levi, Nick Srnicek, and Graham Harman, eds. 2011. *The Speculative Turn: Continental Materialism and Realism*. Melbourne: re:press.

Bull, Michael. 2000. *Sounding Out the City: Personal Stereos and the Management of Everyday Life*. London: Bloomsbury.

Bull, Michael. 2007. *Sound Moves: iPod Culture and Urban Experience*. New York: Routledge.

Bull, Michael, ed. 2013. *Sound Studies*, 4 vols. New York: Routledge.

Chandola, Tripta. 2010. "Listening into Others: In-between Noise and Silence." Ph.D. diss., Queensland University of Technology.

Cimini, Amy. 2011. "Baruch Spinoza and the Matter of Music: Toward a New Practice of Theorizing Musical Bodies." Ph.D. diss., New York University.

Coates, Norma. 2008. "Sound Studies: Missing the (Popular) Music for the Screens?" *Cinema Journal* 48, no. 1: 123–30.

Comaroff, John, and Jean Comaroff. 2011. *Theory from the South: Or, How Euro America Is Evolving toward Africa*. Boulder, CO: Paradigm.

Connell, Raewyn W. 2007. *Southern Theory: Social Science and the Global Dynamics of Knowledge*. London: Polity.

Coole, Diana, and Samantha Frost, eds. 2010. *New Materialisms: Ontology, Agency, and Politics*. Durham, NC: Duke University Press.

Cox, Christoph, and Daniel Warner, eds. 2004. *Audio Culture: Readings in Modern Music*. London: Continuum.

Daughtry, Martin. 2014. *Listening to War: Sound, Music, Trauma, and Survival in Wartime Iraq*. Oxford: Oxford University Press.

Dawson, Ashley, and Brent Hayes Edwards, eds. 2004. "Global Cities South." *Social Text* 81 (Winter), special issue.

DeNora, Tia. 2000. *Music in Everyday Life*. Cambridge: Cambridge University Press.

Derrida, Jacques. [1967] 1976. *Of Grammatology*, trans. Gayatri Chakravorty Spivak. Baltimore: Johns Hopkins University Press.

Santos, Boaventura de Sousa. 2014. *Epistemologies of the South: Justice against Epistemicide*. New York: Routledge.

Eisenberg, Andrew. 2013. "Islam, Sound and Space: Acoustemology and Muslim Citizenship on the Kenyan Coast." In *Music, Sound and Space*, ed. Georgina Born, 186–202. Cambridge: Cambridge University Press.

Erlmann, Veit, ed. 2004. *Hearing Cultures: Essays on Sound, Listening and Modernity*. New York: Berg.

Erlmann, Veit. 2010. *Reason and Resonance: A History of Modern Aurality*. New York: Zone.

Feld, Steven. 1986. "Orality and Consciousness." In *The Oral and the Literate in Music*, ed. Tokumaru Yosihiko and Yamaguti Osamu, 18–28. Tokyo: Academie Music.

Feld, Steven. 1996. "Waterfalls of Song: An Acoustemology of Place Resounding in Bosavi, Papua New Guinea." In *Sense of Place*, ed. Steven Feld and Keith Basso. Santa Fe, NM: School of American Research Press.

Feld, Steven. 2015. "I Hate 'Sound Studies.'" Accessed June 18, 2018. https://static1 .squarespace.com/static/545aad98e4b0f1f9150ad5c3/t/55ef81ffe4b0e4ff182017bf /1441759743397/I+Hate.pdf.

Friedner, Michele, and Stefan Helmreich. 2012. "Sound Studies Meets Deaf Studies." *Senses and Society* 7, no. 1: 72–86.

Gallope, Michael. 2011. "Technicity, Consciousness, and Musical Objects." In *Music and Consciousness: Philosophical, Psychological, and Cultural Perspectives*, ed. David Clarke and Eric Clarke, 47–64. Oxford: Oxford University Press.

Glacken, Clarence J. 1976. *Traces on the Rhodian Shore: Nature and Culture in Western Thought from Ancient Times to the Eighteenth Century*. Berkeley: University of California Press.

Goodman, Steve. 2009. *Sonic Warfare: Sound, Affect, and the Ecology of Fear*. Cambridge, MA: MIT Press.

Gopinath, Sumanth. 2013. *The Ringtone Dialectic: Economy and Cultural Form*. Cambridge, MA: MIT Press.

Gopinath, Sumanth, and Jason Stanyek, eds. 2014. *The Oxford Handbook of Mobile Music Studies*, 2 vols. Oxford: Oxford University Press.

Graeber, David. 2015. "Radical Alterity Is Just Another Way of Saying 'Reality': A Reply to Eduardo Viveiros de Castro." *HAU: Journal of Ethnographic Theory* 5, no. 2. Accessed June 18, 2018. https://www.haujournal.org/index.php/hau/article/view /hau5.2.003/1978.

Hilmes, Michele. 2005. "Is There a Field Called Sound Culture Studies? And Does It Matter?" *American Quarterly* 57, no. 1: 249–59.

Hilmes, Michele. 2008. "Foregrounding Sound: New (and Old) Directions in Sound Studies." *Cinema Journal* 48, no. 1: 115–17.

Holbraad, Martin, and Morten Axel Pedersen. 2017. *The Ontological Turn: An Anthropological Exposition*. Cambridge: Cambridge University Press.

Holbraad, Martin, Morten Axel Pedersen, and Eduardo Viveiros de Castro. 2013. "The Politics of Ontology: Anthropological Positions." Position paper for a roundtable discussion at the Annual Conference of the American Anthropological Association, Chicago, November 21.

Ingold, Tim. 2000. *The Perception of the Environment: Essays on Livelihood, Dwelling, and Skill*. London: Routledge.

Irving, David. 2014. "Hybridity and Harmony: Nineteenth-Century British Discourse on Syncretism and Intercultural Compatibility in Malay Music." *Indonesia and the Malay World* 42, no. 123: 197–221.

Kahn, Douglas. 2001. *Noise, Water, Meat: A History of Voice, Sound, and Aurality in the Arts*. Cambridge, MA: MIT Press.

Kane, Brian. 2015. "Sound Studies without Auditory Culture: A Critique of the Ontological Turn." *Sound Studies* 1, no. 1: 2–21.

Kassabian, Anahid. 2013. *Ubiquitous Listening: Affect, Attention, and Distributed Subjectivity*. Berkeley: University of California Press.

Kelly, Caleb, ed. 2011. *Sound*. Cambridge, MA: MIT Press.

Kim-Cohen, Seth. 2009. *In the Blink of an Ear: Toward a Non-Cochlear Sonic Art*. London: Bloomsbury.

Kirsky, Eben, ed. 2014. *The Multispecies Salon*. Durham, NC: Duke University Press.

Kreutzer, Tino. 2009. "Accessing Cell Phone Usage in a South African Township School." *International Journal of Education and Development* 5, no. 5. Accessed June 18, 2018. http://ijedict.dec.uwi.edu/viewarticle.php?id=862&layout=html.

Kunreuther, Laura. 2014. *Voicing Subjects: Public Intimacy and Mediation in Kathmandu*. Stanford, CA: University of California Press.

Kusno, Abidin. 2010. *The Appearances of Memory: Mnemonic Practices of Architecture and Urban Form in Indonesia*. Durham, NC: Duke University Press.

LaBelle, Brandon. 2010. *Acoustic Territories: Sound Culture and Everyday Life*. New York: Continuum.

Larkin, Brian. 2004. "Degraded Images, Distorted Sounds: Nigerian Video and the Infrastructure of Piracy." *Public Culture* 16, no. 2: 289–314.

Latimer, Quinn, and Adam Szymczyk, eds. 2015. *South as a State of Mind: Documenta 14*. Cologne: Walther König.

Latour, Bruno. 1993. *We Have Never Been Modern*, trans. Catherine Porter. Cambridge, MA: Harvard University Press.

Livermon, Xavier. 2008. "Sounds in the City." In *Johannesburg: The Elusive Metropolis*, ed. Sarah Nuttall and Achille Mbembe, 271–84. Durham, NC: Duke University Press.

Mahler, Anne Garland. 2015. "The Global South in the Belly of the Beast: Viewing African American Civil Rights through a Tricontinental Lens." *Latin American Research Review* 50, no. 1: 95–116.

Meillassoux, Quentin. 2008. *After Finitude: An Essay on the Necessity of Contingency*, trans. Ray Brassier. London: Continuum.

Minh-ha, Trinh T. 1989. *Woman, Native, Other: Writing Postcoloniality and Feminism*. Bloomington: Indiana University Press.

Morat, Daniel. 2014. *Sounds of Modern History: Auditory Cultures of Europe*. Oxford: Berghahn.

Moreno, Jairo. 2013. "On the Ethics of the Unspeakable." In *Speaking of Music: Addressing the Sonorous*, ed. Keith Chapin and Andrew H. Clark, 212–41 New York: Fordham University Press.

Morris, Rosalind. 2008. "The Miner's Ear." *Transition* 98: 96–115.

Morris, Rosalind. 2010. "Accident Histories, Post-Historical Practice? Re-reading *Body of Power, Spirit of Resistance* in the Actuarial Age." *Anthropological Quarterly* 83, no. 3: 581–642.

Motlatsi. 2016. "In Conversation with Milisuthando Bongela: Black Consciousness through a Critical Art and the Art of the Critical Consciousness." *Bubblegum*. Accessed June 10, 2017. http://www.bubblegumclub.co.za/conversation-milisuthando -bongela-black-consciousness-critical-art-art-critical-consciousness.

Muller, Carol Ann, and Sathima Benjamin. 2011. *Musical Echoes: South African Women Thinking in Jazz*. Durham, NC: Duke University Press.

Napolin, Julie Beth. 2013. "A Sinister Resonance: Vibration, Sound, and the Birth of Conrad's Marlow." *Qui Parle* 21, no. 2: 69–100.

Novak, David. 2011. "The Sublime Frequencies of New Old Media." *Public Culture* 23, no. 3 (65): 603–34.

Novak, David, and Matt Sakakeeny, eds. 2015. *Keywords in Sound*. Durham, NC: Duke University Press.

Nuttall, Sarah, and Achille Mbembe, eds. 2008. *Johannesburg: The Elusive Metropolis*. Durham, NC: Duke University Press.

Nyong'o, Tavia. 2014. "Afro-Philo-Sonic Fictions: Black Sound Studies after the Millennium." *Small Axe* 18, no. 2 (44): 173–79.

Obbario, Juan. 2012. "Symposium: Theory from the South." *Salon* 5: 5–36.

Ochoa Gautier, Ana María. 2006. "Sonic Transculturation, Epistemologies of Purification and the Aural Public Sphere in Latin America." *Social Identities* 12, no. 6: 803–25.

Ochoa Gautier, Ana María. 2014. *Aurality: Listening and Knowledge in Nineteenth-Century Colombia*. Durham, NC: Duke University Press.

Pinch, Trevor, and Karin Bijsterveld, eds. 2011. *The Oxford Handbook of Sound Studies*. Oxford: Oxford University Press.

Porcello, Thomas, Louise Meintjes, Ana María Ochoa, and David W. Samuels. 2010. "The Reorganization of the Sensory World." *Annual Review of Anthropology* 39: 51–66.

Povinelli, Elizabeth. 2016. *Geontologies: A Requiem to Late Liberalism*. Durham, NC: Duke University Press.

Radano, Ronald. 2016. "Black Music Labor and the Animated Properties of Slave Sound." *boundary 2* 43, no. 1: 173–208.

Radano, Ronald, and Tejumola Olaniyan, eds. 2016. *Audible Empire: Music, Global Politics, Critique.* Durham, NC: Duke University Press.

Rasmussen, Anne. 2010. *Women, the Recited Qur'an, and Islamic Music in Indonesia.* Berkeley: University of California Press.

Reddy, Gayatri. 2005. *With Respect to Sex: Negotiating Hijra Identity in South India.* Chicago: University of Chicago Press.

Rousseau, Jean-Jacques. 1781. *Essai sur l'origine des langues.* Geneva, Switzerland.

Rowell, Lewis. [1992] 2015. *Music and Musical Thought in Early India.* Chicago: University of Chicago Press.

Roy, Ananya, and Emma Shaw Crane, eds. 2005. *Territories of Poverty: Rethinking North and South.* Athens: University of Georgia Press.

Samuels, David W., Louise Meintjes, Ana María Ochoa, and Thomas Porcello. 2010. "Soundscapes: Toward a Sounded Anthropology." *Annual Review of Anthropology* 39: 329–45.

Sartori, Andrew. 2008. *Bengal in Global Concept History: Culturalism in the Age of Capital.* Chicago: University of Chicago Press.

Sartori, Andrew. 2014. *Liberalism in Empire: An Alternative History.* Berkeley: University of California Press.

Schafer, R. Murray. 1977. *The Tuning of the World.* New York: Knopf.

Schmidt, Leigh Eric. 2002. *Hearing Things: Religion, Illusion, and the American Enlightenment.* Cambridge, MA: Harvard University Press.

Simone, AbdouMaliq. 2009. *City Life from Jakarta to Dakar: Movements at the Crossroads.* New York: Routledge.

Stadler, Gustavus, ed. 2010. "Never Heard Such a Thing: Lynching and Phonographic Modernity." *Social Text* 28, no. 1: 87–105.

Starobinski, Jean. 1989. *La reméde dans le mal.* Paris: Gallimard.

Steingo, Gavin. 2015. "Sound and Circulation: Immobility and Obduracy in South African Electronic Music." *Ethnomusicology Forum* 24, no. 1: 102–23.

Steingo, Gavin. 2016. *Kwaito's Promise: Music and the Aesthetics of Freedom in South Africa.* Chicago: University of Chicago Press.

Sterne, Jonathan. 2003. *The Audible Past: Cultural Origins of Sound Reproduction.* Durham, NC: Duke University Press.

Sterne, Jonathan. 2012a. "Sonic Imaginations." In *The Sound Studies Reader,* ed. Jonathan Sterne, 1–17. New York: Routledge.

Sterne, Jonathan, ed. 2012b. *The Sound Studies Reader.* New York: Routledge.

Sterne, Jonathan. 2015. "Hearing." In *Keywords in Sound,* ed. David Novak and Matt Sakakeeny, 65–77. Durham, NC: Duke University Press.

Stiegler, Bernard. 1998. *Technics and Time, I: The Fault of Epimetheus,* trans. Richard Beardsworth and George Collins. Stanford, CA: Stanford University Press.

Stoever, Jennifer Lynn. 2016. *The Sonic Color Line: Race and the Cultural Politics of Listening*. Durham, NC: Duke University Press.

Sykes, Jim. 2015. "Sound, Religion, and Public Space: Tamil Music and the Ethical Life in Singapore." *Ethnomusicology Forum* 24, no. 3: 485–513.

Sykes, Jim. 2017. "Sound as Promise and Threat: Drumming, Collective Violence, and the British Raj in Colonial Ceylon." In *Cultural Histories of Sound and Listening in Europe, 1300–1918*, ed. Ian Biddle and Kirsten Gison. New York: Routledge.

Sykes, Jim. 2018. *The Musical Gift: Sonic Generosity in Post-War Sri Lanka*. New York: Oxford University Press.

Szendy, Peter. 2008. *Listen: A History of Our Ears*. New York: Fordham University Press.

Szendy, Peter. 2015. "The Auditory Re-Turn (The Point of Listening)." In *Thresholds of Listening: Sound, Technics, Space*, ed. Sander van Maas, 18–29. New York: Fordham University Press.

Tausig, Benjamin. 2013. "Review Article" [Georgina Born, ed., *Music, Sound and Space*; Jonathan Sterne, ed., *The Sound Studies Reader*; and Trevor Pinch and Karin Bijsterved, *The Oxford Handbook of Sound Studies* ]. *Twentieth-Century Music* 11, no. 1: 163–76.

Thomas, Downing A. 1995. *Music and the Origins of Language: Theories from the French Enlightenment*. Cambridge: Cambridge University Press.

Thompson, Emily. 2002. *The Soundscape of Modernity: Architectural Acoustics and the Culture of Listening in America, 1900–1933*. Cambridge, MA: MIT Press.

Tien, Adrian, 2015. *The Semantics of Chinese Music: Analysing Selected Chinese Musical Concepts*. Amsterdam: John Benjamins.

Tomlinson, Gary. 2007. *The Singing of the New World: Indigenous Voice in the Era of European Contact*. Cambridge: Cambridge University Press.

Tsing, Anna. 2015. *The Mushroom at the End of the World: On the Possibility of Life in Capitalist Ruins*. Princeton, NJ: Princeton University Press.

Van Maas, Sander, ed. 2015. *Thresholds of Listening: Sound, Technics, Space*. New York: Fordham University Press.

Vanzolini, Marina, and Pedro Cesarino. 2014. "Perspectivism." *Oxford Bibliographies Online*. Accessed June 18, 2018. doi: 10.1093/OBO/9780199766567-0083.

Venkatesan, Soumhya, ed. 2010. "Ontology Is Just Another Word for Culture: Motion Tabled at the 2008 Meeting of the Group for Debates in Anthropological Theory, University of Manchester." *Critique of Anthropology* 30, no. 2: 152–200.

Vergès, Françoise. 2015. "Like a Riot: The Politics of Forgetfulness, Relearning the South, and the Island of Dr. Moreau." In *South as a State of Mind 6*, ed. Quinn Latimer and Adam Szymczyk. *Documenta* 14, no. 1 (fall–winter): 26–43.

Viveiros de Castro, Eduardo. 1998. "Cosmological Deixis and Amerindian Perspectivism." *Journal of the Royal Anthropological Association* 4: 469–88.

Viveiros de Castro, Eduardo. 2003. "And." *Manchester Papers in Social Anthropology* 7. Accessed June 18, 2018. https://sites.google.com/a/abaetenet.net/nansi/abaetextos/anthropology-and-science-e-viveiros-de-castro.

Viveiros de Castro, Eduardo. 2004. "Perspectival Anthropology and the Method of Controlled Equivocation." *Tipití* 2, no. 1: 3–22.

Viveiros de Castro, Eduardo. 2015. "Who's Afraid of the Ontological Wolf: Some Comments on an Ongoing Anthropological Debate." *Cambridge Anthropology* 33, no. 1: 2–17.

Weheliye, Alexander. 2000. "In the Mix: Hearing the Souls of Black Folk." *Amerikastudien/American Studies* 45, no. 4: 535–54.

Weheliye, Alexander. 2014. "Engendering Phonographies: Sonic Technologies of Blackness." *Small Axe* 18, no. 2(44): 180–90.

Weidman, Amanda. 2006. *Singing the Classical, Voicing the Modern: The Postcolonial Politics of Music in South India*. Durham, NC: Duke University Press.

White, Shane, and Graham White. 2006. *The Sounds of Slavery: Discovering African American History through Songs, Sermons, and Speech*. Boston: Beacon.

Yuran, Noam. 2014. *What Does Money Want? An Economy of Desire*. Stanford, CA: Stanford University Press.

# [ PART I ]

## THE TECHNOLOGY PROBLEMATIC

.....

# Another Resonance

## AFRICA AND THE STUDY OF SOUND

*Gavin Steingo*

.....

## PROLEGOMENA

Heading south from the suburbs of Sandton to the townships of Soweto, South Africa, one crosses a threshold of automobile habitation. In Sandton, car windows are generally closed to avoid the heckling of roadside vendors and to create an acoustic environment of one's own desire. When entering Soweto, my local friends insist, a driver must roll down his or her windows, unbuckle the seat belt, and cruise at a steadily slow pace, especially because closed windows for an outsider (and a white person) indicate anxiety and arouse suspicion. Wide-open windows — paradoxically — decrease vulnerability, signaling a familiarity with township space, on the one hand, and a non-neurotic openness to "strangers," on the others. With windows open and pedestrians constantly brushing by the sides of the vehicle, a car's boundaries become porous. Intimacies are established through sonic exchanges (greetings, whistles, shouts) between drivers and others who inhabit the township road — pedestrians, horse-cart riders, mobile fruit vendors, and herd boys, for example.

In many central African towns and cities — such as the town of Limbe, Cameroon, where I have also conducted fieldwork — motorcycle taxis carrying as many as four passengers whiz past workers traveling in the uncovered rear beds of pickup trucks, while women carrying buckets of grain on their heads call out to young boys stealthily darting between quickly moving vehicles. Having departed Limbe toward the dense rainforest region in the south, I notice one day how the driver of a shared taxi sedan mutes the sound system that has been blasting intricate weaving guitar lines of *bikutsi* for the past several hours. As the driver approaches each sharp turn on the muddy forest road, he honks loudly to alert the occasional car moving in the opposite direction, and because he expects other drivers to do the same, he listens carefully, cautiously, for the remainder of our trip.

Motor vehicles are crucial technologies of mobile sound. But they are not, of course, the only ones. Back in Soweto, I notice how digital formats such as MP3s are transferred through memory sticks or hard drives, but hardly ever online. I make a plan to follow a single MP3—perhaps a short digital music composition—as it moves from a local producer through a network of friends and acquaintances. But I quickly realize that this plan is hopeless, because the digital storage devices used by my interlocutors keep breaking, getting lost (or stolen?), being misplaced.

On a trip to Cameroon in 2008, I observe a practice closely resembling one in Mali that will make international headlines seven years later. Writing for the *New York Times*, Lydia Polgreen (2015) describes *téléchargeurs* (or downloaders), who "operate as an offline version of iTunes, Spotify and Pandora all rolled into one." Téléchargeurs download large playlists, then transfer songs to consumers for a small fee via "memory cards or USB sticks, or directly onto cellphones" (Polgreen 2015). In such scenarios, music distribution oscillates between online and offline infrastructures, variously hopping or creeping, depending on the technology employed. In Cameroon, I had occasion to interact with downloaders fulfilling a role identical to téléchargeurs, as well as with young men and boys selling cheap pirated compilation CDs. (At an Internet café down the road from the abandoned monastery where I was staying, I encountered groups of young men scamming international buyers by selling purebred poodles that do not exist.)

Another aspect I have long observed during fieldwork in Africa is the fragility of technology and the concomitant (although obviously much more serious) insecurity of human life. Although there is no need to resort to theoretical extravagances such as "bare life," it is certainly the case that in many parts of Africa human bodies are forced into spaces of extreme vulnerability. In South Africa, black bodies—particularly the bodies of black women—are vulnerable to multiple forms of violence and injury. Most relevant for this chapter are injuries to the human sensorium, to those organs of sensation (such as the ear) through which humans encounter the world. I have more to say about this later.

.........

*Gavin Steingo*

The varied auditory experiences and modalities of sound production mentioned in the preceding ethnographic vignettes are almost entirely absent from sound studies. Precisely why this is so, and what such a striking lacuna may mean for an emerging discipline, is addressed throughout the course of this chapter. In what follows, I stage a series of dialogical encounters between recent sound studies literature and my ethnographic fieldwork in Africa—primarily South Africa and Cameroon. I respond to three major claims of sound studies, as noted in the introduction: (1) that technology is increasingly isolating listening subjects into individual "bubbles"—for example, in automobiles and through mobile listening devices; (2) that music has become increasingly available and ubiquitous due to technological advances in circulation; and (3) that listening is associated with biopolitical investment and efficiency. The ethnographic vignettes already suggest that these claims are extremely limited. In many parts of Africa, there exists a radically different relationship to sound. Listening carefully for these other relationships goes a long way toward "remapping" sound studies. It also challenges us to rethink the meaning of modernity itself.

## SOUND STUDIES: THREE KEY CLAIMS

In this section, I engage the three claims outlined earlier by elaborating their intellectual history with reference to particular authors. While the following discussion is by no means comprehensive, it maps out the broad strokes of some of the primary preoccupations within sound studies.

### Contemporary Listening: Mobile and Private

Scholars of auditory culture have established a remarkably consistent, if limited, narrative of urban listening that typically revolves around two interrelated terms: "mobility" and "privatization." While often referring back to a stable set of early- and mid-twentieth-century theorizations—notably, Georg Simmel's ([1903] 1997) writing on atomization, Siegfried Kracauer's ([1927] 1995) work on isolation and alienation, and Theodor Adorno's ([1927] 2002) critique of recorded music—it was perhaps Raymond Williams's notion of "mobile privatization" that most clearly set the tone for work that followed. For Williams, the advent of television in the 1920s marked a turning point in media history. In *Television: Technology and Cultural Form*, Williams (1974: 20) observed about North America and England in the 1920s and '30s: "The earlier period of public technology, best exemplified by the railways and city lighting, was being replaced by a kind of technology for which no satisfactory name has yet to be found: that which served an at once mobile and home-centered way of living: a form of *mobile privatisation.*"

In the decades that followed Williams's *Television*, several theorists came to similar conclusions, although from decidedly dissimilar vantage points. Of particular relevance is Paul Virilio's brief dystopian text, "The Last Vehicle," which describes the "audiovisual vehicle[s]" of electronic media as static transportation devices characterized by "ecstasy, music, and speed" (Virilio 1989: 115). According to Virilio, this vehicle "ought at last to bring about the victory of sedentariness, this time an ultimate sedentariness" (109). For Virilio, like Williams before him, the "mobility" of electronic media is metaphorical: televisions bring the flux of an outside world to a stationary viewer located in private, domestic space.

But despite Virilio's anxious warning, cultural critics began to observe a tendency *away* from the virtual travel of sedentary listening and toward new practices of listening on the move. In the early 1980s, Shūhei Hosakawa described the so-called Walkman effect, the effect of a literal mobile listening and thus the "*autonomy-of-walking-itself*" (Hosakawa 1984: 166). Hosakawa presented a four-phase history of "*musica mobilis*," with the Walkman listener coming to represent the final stage.[1] Saturated with Deleuzian terminology, Hosakawa's "The Walkman Effect" characterizes this final stage in terms of miniaturization (a capacity for increased mobility), singularization (a set of impersonal and nomadic "emissions"), autonomy ("a radically *positive distance*"), and meaning construction (in particular, a generalization of the surface).

Interest in mobile listening continued strongly into the 1990s. One author wrote about the "mobile, wraparound world" of the Sony Walkman, that "serves to set one apart while simultaneously reaffirming individual contact to certain common, if shifting measures (music, fashion, metropolitan life . . . and their particular cycles of mortality)" (Chambers 1990: 2). Paul du Gay and the other authors of *Doing Cultural Studies: The Story of the Sony Walkman* similarly point to the manner by which mobile listening troubles the boundary between the individual and the common, the private and public. In their view, the Walkman marks a rupture with the past because it takes "*private listening into the public domain*" (du Gay et al. 1997: 106).

These early texts gave way, in the twenty-first century, to a proliferation of studies on mobile listening, with an emphasis on automobiles, on the one hand, and newer mobile listening devices (such as the iPod), on the other. More than ever before, these studies explore the border between the enclosed

*Gavin Steingo*

space of listening and the world perceived as exterior. Michael Bull (2001: 358), for example, "discusses the manner in which we lay claim to the spaces we inhabit through the automobile," where the automobile is taken to be a "metaphor for the dominant values of individualism and private property." Drawing liberally from journalistic accounts, critical theory, and interviews with drivers, he establishes an understanding of automobile listening as a form sonic enclosure that mediates experiences with the world outside the vehicle. He suggests that the auditory space of the car is a mobile "bubble" (Bull 2001: 195, after Baudrillard 1993), a "sonic envelop" (Bull 2004: 247), and a "physical cocoon" (Bull 2001: 358).[2]

Karin Bijsterveld (2010: 192), similarly, investigates how and why cars were turned into "acoustic cocoons, that is, into domains in which people experience privacy and relaxation because they consider the interior acoustics of cars pleasant and controllable." And, to provide one last representative example, Brandon LaBelle (2010: 142) argues that the automobile "lends greatly to personalizing movements through the world, cocooning itself within a stylized interior space. . . . A living room on wheels, a sonic bubble, the car is a total design giving enormous power to the driver."[3] According to LaBelle (2010), the car is somewhat unique in its coupling of auditory enclosure with the "controlling [of] public presence." He concludes, "The car is a *second skin*" (143).

Bull's later work on the iPod follows the same threads: he is interested, ultimately, in how people manage their daily activities through private listening. The car and the mobile listening device therefore represent two forms of mobile and private listening, two modes of cocooning, enveloping, or "enbubbling" an individual with sound. Even more so than his other research, Bull's work on iPod listening is striking for its level of cultural specificity. His respondents include "a 32-year-old publisher," "a 23-year-old systems analyst," "a 35-year-old bank manager," and "a 31-year-old IT specialist," among others (Bull 2005: 354n1). Consider how culturally specific the following words are, uttered by Joey, a twenty-eight-year-old researcher living in New York:

> When I leave my apartment in the morning I grab my iPod and shove it in my pocket. By the time I get to the subway platform I am listening to my morning mix. This mix is 80s music ranging from Eurythmics to Blondie and The Smiths. It's an upbeat and a subtle mix that wakes me up and gets me motivated for my day. I will admit that some days I am not into the mood to go to work so I will put on something more sombre like Cat Power. (quoted in Bull 2005: 348)

As hip as Joey's mix may have been in the early 2000s, the entire scenario that he paints (grabbing his iPod, waiting on the subway platform, listening to retro '80s tunes, switching to Cat Power's somber sound world) would be completely alien to most music listeners in Johannesburg and Lagos, in Jakarta and Mumbai.

In fact, the histories and modalities of listening described by all of the authors covered in this section are resolutely Northern. In the best cases, this orientation is made explicit from the outset. Bull (2007: 2, emphasis added), for example, writes, "The Gothic cathedral to [the] Citroën DS to the Apple iPod represents a *Western narrative of increasing mobility and privatization*," while Bijsterveld, in an extended cowritten work on auditory cocooning, notes that her "empirical research is focused on Western Europe and North America" (Bijsterveld et al. 2014: 2). On the one hand, these writers simply limit their research to Northern contexts. On the other hand, however, it seems to me that their specific claims occasionally lapse into universalism. How else can we interpret a statement such as, "iPod culture represents a world in which we all possess mobile phones, iPods or automobiles—it is a culture which universalizes the privatization of public space, and it is a largely auditory privatization"? (Bull 2007: 4). At the risk of being overly pedantic, it is worth questioning who the "we" is in such an account. When Bull (2004: 243, emphasis added) writes, "An increasing number of *us* demand the intoxicating mixture of noise, proximity and privacy while on the move," and when Bijsterveld (2010: 201, emphasis added) observes, "A separate research industry has emerged that studies what *you and I* consider pleasant sounds," then we—as critical scholars—need to question who the "us" and the "you and I" refer to. As becomes very clear when reading *Sound and Safe: A History of Listening behind the Wheel* (Bijsterveld et al. 2014), the automobile research industry certainly was not especially interested in what residents of Soweto or Limbe consider pleasant sounds.

### Modern Sound: Availability and Ubiquity

A second and related corpus of literature points to music's ever increasing availability and ubiquity. In popular media publications, the infinite availability of all music is often referred to as a "celestial jukebox."[4] As one author writes, "The ultimate goal for music technology, 'the celestial jukebox,' is going to be reached very soon" (Wolk 2009). Another announces more dramatically in the title of his article that the "Celestial Jukebox Falls to Earth" (Van Buskirk 2006). Academics are often surprisingly close in their pronouncements. Lars Holmquist (2005: 71), for example, states baldly, "A combination of growing

*Gavin Steingo*

disc capacities, compression algorithms, and increasing bandwidth means we can have an almost limitless supply of music just about everywhere. By replacing corporate Muzak and conservative radio schedules with portable MP3 players, online music stores, file sharing, ringtone downloads, and celebrity playlists, we herald the age of ubiquitous music."

More recently, authors have essentially argued for the actual existence of the celestial jukebox under another name: the "cloud." As Jeremy Morris (2001) writes in an article on cloud-based music services, "The cloud offers an infinite space where music is ever available."[5] As these examples illustrate, music's ubiquity and availability is usually attributed to advances in technology. Gil Weinberg (2003: 3) affirms, "Music today is more ubiquitous, accessible, and democratized than ever. Thanks to technologies such as high-end home studios, audio compression, and digital distribution, music now surrounds us in everyday life, almost every piece of music is a few minutes of download away, and almost any western musician, novice or expert, can compose, perform and distribute their music directly to their listeners from their home studios."

As was the case with the previous section on mobile and private listening, here it is important once again to ask: to whom does this writing apply? Holmquist employs that most subtle and troubling word—"we"—while in Morris's work perspective is simply absent (the cloud simply "offers an infinite space"). Weinberg, meanwhile, presents the ubiquity, accessibility, and democratization of music in nonspecific terms, only to inject a qualifier halfway through one sentence—"and almost any *western* musician."

It is also worth briefly noting the caveat on the first page of Anahid Kassabian's celebrated *Ubiquitous Listening*, which is very relevant to my own research and which I find quite instructive. After stating boldly that "our days are filled with listening," Kassabian (2013: xi) immediately anticipates the rebuttal: "Of course, you will object: some people more than others, some countries more than others, some economies more than others." And because such objections are indeed warranted, it is worthwhile to take note of Kassabian's response. She writes: "But a colleague told me that he heard music in a supermarket on a dirt road in South Africa, so let us not leap to conclusions about the lives of others."[6] With the matter firmly resolved, she concludes that she "will be happy to think about England the United States."

This passage takes the form of a four-part argument in which Kassabian (1) states her argument ("our days are filled with listening"); (2) notes global unevenness ("some countries more than others, some economies more than others"); (3) implies that this unevenness is becoming flattened with global convergence ("a colleague told me that he heard music in a supermarket on a dirt road in South Africa"); and (4) returns "happily" to the global North, only to repeat—with complete fidelity—her original argument. But this line of argumentation does not seriously consider the global South, since a casual remark by a colleague surely cannot count as evidence. It seems to me that Kassabian glosses over global difference far too quickly.

To be sure, recent efforts in sound studies have presented a more balanced and critical perspective. Sumanth Gopinath and Jason Stanyek's introduction to the first volume of the *Oxford Handbook of Mobile Music Studies* is particularly relevant in this regard. While noting that "billions of people take for granted the perpetual access to recorded music" (Gopinath and Stanyek 2014: ix), they also observe that music and sound have never been "tethered to specific spaces and places" (2). Gopinath and Stanyek (15) make the crucial point that the notion of "anywhere, anytime" music describes an *ideology* just as much as—perhaps even more so than—a material reality. Indeed, "anytime, anywhere" has "operated as a key advertising trope for the mobile music industry" for more than a century (15).

But even considering their careful analysis of musical ubiquity, Gopinath and Stanyek are strangely silent on what this ideology means for most of the world. Almost all of their examples are taken from North America or Europe. Thus, they cannot present any real alternative to the "anytime, anywhere" ideology that they critique.[7]

### Hearing Loss and Hearing Aids: Biopolitics and Efficiency

The third core argument regarding sound and listening is somewhat less straightforward; it concerns a biopolitical investment in the human sensorium. Such an investment is especially evident in histories of industrial labor. Consider, for example, the context of the early twentieth-century factory floor in Europe and the United States, where noise became a major concern. In 1913, the American Josephine Goldmark, chair of the Committee on the Legal Defense of Labor Laws, stated that noise "not only distracts attention but necessitates a greater exertion of intensity or conscious application, thereby hastening the onset of fatigue of the attention" (quoted in Bijsterveld 2006: 325). As Bijsterveld observes, Goldmark's claim was followed by a number of experiments that seemed to corroborate her argument. Unlike the periodic rhythms of preindustrial labor, the factory was understood as a space of irregular bursts; thus, "enhancing employees'

*Gavin Steingo*

efficiency by restoring the sensation of rhythm on the shop-floor . . . became one of the strategies used to reduce the negative effects of noise"
(Bijsterveld 2006: 326).

Hearing loss was another noise-related threat to productivity. While in
the early twentieth century this problem was dealt with through attempts to
quiet machinery, by mid-century the prophylactic approach to hearing loss
had shifted to earplugs. The aggressive introduction of earplugs in the Unites
States is attributable, in large part, to a "rapid growth in workers' compensation claims" (Bijsterveld 2006: 328).

Bijsterveld masterfully shows that one major factor in hearing loss was
a resistance to earplugs by the workers themselves. Some workers felt the
muffling of machine noise was a *problem*, since the sound of a machine was
often an indicator of its successful functioning or, alternatively, of a technical problem. For others, moderately loud machinery made it possible to
sing during work without disturbing fellow workers (Bijsterveld 2006: 331).
And there were other reasons still: "A study published in 1960 showed that
hearing protection made workers feel insecure about the direction from
which sounds came, and caused communication problems as well as 'a nasty
feeling.' The workers also reported that wearing earplugs meant 'a kind of
embarrassment' in relation to their 'comrades'" (Bijsterveld 2006: 331). By di-
minishing the full spectrum of factory sounds, earplugs deprived workers of
their established labor *culture*. Workers were prepared to experience partial
hearing loss, because the alternative—wearing earplugs—was far worse.

The adoption of ear protection devices has thus been uneven—and so, too,
we might add, has the development of hearing aids. Hearing aids are essentially "mobile listening devices," and as such, their development is intimately
connected to technologies such as the Walkman. While miniaturization is a
trope of many mobile listening technologies, historians of hearing loss have
recently challenged the common narrative of miniaturization as one of "relentless progress" (Mills 2011: 26).[8] While miniaturization has in fact been
a major factor in the development of hearing aids—notably, because users
typically desire devices that are not noticeable to others—smaller devices
historically have also meant less effective functioning. Mara Mills (2011: 26)
thus concludes, "Consumers have always been forced to weigh trade-offs
among cost, function, size, and style."

Despite the different contexts that Bijsterveld and Mills examine, both
present a picture consistent with Michel Foucault's classic formulation of
biopower: to "'make' live and 'let' die" (Foucault 2003: 241). For the factory

workers described by Bijsterveld, there is an investment in vitality and productivity. Although this investment occasionally accepts moderate hearing
loss, the acceptance is always in the service of a more general calculus of
optimization. By *not* hearing well, workers are able to enjoy their own singing
voices and perform a kind of masculine ruggedness. Here, "making live"
remains the primary organizing principle, even if it requires a trade-off in
which hearing is partially sacrificed. The hard-of-hearing individuals described by Mills likewise choose between hearing better and appearing more
"normal" to others. Either way, the calculus is one of general optimization
for the subject.

### SOUTHERN RESONANCES

None of the three major claims of sound studies meaningfully applies to
most African contexts. Recalling briefly the ethnographic vignettes with
which this chapter began, it is obvious that in the African contexts described,
automobiles are in no way sonic "bubbles," "envelops," or "cocoons." A driver
in Soweto *must*—for both safety and courtesy—open his or her windows
and participate in a number of sonic exchanges with a host of fellow travelers
on the road. There, it is pedestrians and not drivers who are warned about
alcohol consumption—as the carton of a locally-produced alcoholic drink
warns: "DON'T DRINK AND WALK ON THE ROAD, YOU MAY BE KILLED."[9]
On the streets of Limbe, the boundary between interior and exterior is extremely porous: motorcycles and pickup trucks are precariously open to the
outside. In rainforest areas, moreover, taxi drivers open windows and navigate the road by listening for oncoming vehicles whose approach is visibly
obscured by thick vegetation and winding streets.

The claims about musical availability and ubiquity also do not match
most African contexts. In Soweto, the MP3 is not primarily an online format
and, as such, MP3 move slowly and awkwardly. In Cameroon and Mali, téléchargeurs, or downloaders, do access that "infinite space where music is ever
available" (as Jeremy Morris calls it), but the ordinary music consumer relies
precisely on these downloaders for access to the cloud. As James Ferguson
has pointed out, most things do not simply "flow." (He is talking about capital, but the same holds true for music stored as digital information.) Para-
phrasing Ferguson (2006: 38, emphasis added), we can say of music that it
does not flow as much as "it *hops*, neatly skipping over much of what lies in
between." In Mali, music hops directly to the téléchargeurs and subsequently
creeps from person to person through Bluetooth or the exchange of physical

*Gavin Steingo*

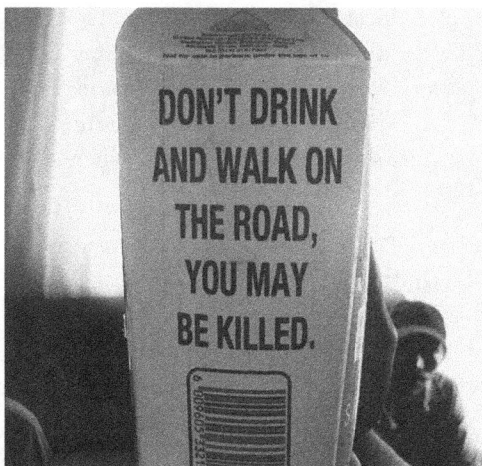

FIG 1.1. An iJuba carton at a Soweto tavern, 2012. Photograph by the author.

hard drives. Instead of an unlimited availability and ubiquity of music, one finds an uneven topography with "jagged edges" (Ferguson 2006: 48).

Moving finally to the third major claim, I would argue that in many parts of the world there is little biopolitical investment in citizens. As Rosalind Morris (2008: 111) notes in a brilliant text about gold mining in twentieth-century South Africa, white mine bosses essentially ignored hearing loss in black miners because a disabled miner was simply "abandoned to fate," and a new, young miner was recruited to replace him. Hence, in South Africa the discussions about earplugs, workers' compensation, and ungainly hearing aids are out of place.

Stated bluntly, when one moves out of a quite narrow geographic focus, claims and theories quickly fall apart.

Of course, not all researchers writing about sound promulgate the three claims I examine in this chapter. Even scholars working on Euro-American listening have begun to nuance some of the arguments made in seminal sound studies texts. David Beer (2007: 857) suggests that Bull's work on mobile listening reflects less the empirical reality of listeners than "a *utopian zone of exclusion* formed around the user." Beer makes a very simple and elegant observation about iPod listeners: "Sound generated in their ears by the mobile music device is not the only sound they hear: they are still exposed to soundscapes of the urban territories through which they pass, and in fact, if the earphones are loud enough, may also be contributing to other people's experiences through sound leakage" (Beer 2007: 858). In other words, the iPod (or iPhone or Walkman) listener does walk around in a sonic bubble.

In his stunning ethnography of children's listening practices at a school in the U.S. state of Vermont, Tyler Bickford makes an even more profound observation: children use MP3 players such as iPods *not* as private listening devices but, rather, as social objects. To better elucidate how this is the case, it is not possible to improve on Bickford's poignant description, which I here quote at length:

> I saw eighth-graders Amber and Daisy sitting side-by-side on the swings, talking in a group of friends while listening to Amber's iPod, the earbud cables stretched across the eighteen inches between the swings. I was impressed that they so easily shared the earbuds even as they swayed back and forth, and I asked if they would ever listen together and swing at the same time. They took my question as a challenge, and Daisy turned to Amber with a mischievous look as they began to pump their legs, almost hitting their friends who scrambled out of the way. They laughed and cheered each other on as they swung higher and higher, coordinating their leg pumps to stay connected by the precariously balanced iPod earbuds in their ears. They swung together like that until they couldn't go any higher, and the earbud only dropped out of Daisy's ear when they finally fell out of sync while slowing down from the peak of their swing. When they came to a stop Daisy looked at me, pleased and defiant, demanding acknowledgment of their physical accomplishment: "See?" (Bickford 2014: 336)

Quite unlike the solitary listeners that populate much recent literature, for these children "earbuds are good for sharing."[10]

A quite different take on headphone listening can be found in Rey Chow's pioneering article "Listening Otherwise, Music Miniaturized" (1990–91), about the politics of sound in East Asia. While couched within a larger argument about what it means to have a voice (and why this question seems important in the first place), one section of Chow's text in particular offers a robust alternative to the hegemonic narrative of private mobile listening to which we have become accustomed.

Like most writers on mobile listening (Beer and Bickford being the exceptions), Chow (1990–91: 145) begins by observing that the Walkman is essentially antisocial, creating "a blockage between 'me' and the world." But in the context of communist China, the "social" or the "collective" means something very different. As she writes, "It is when we deal with the Third World that we have to be particularly careful in resorting to paradigms of the collective as such. Why? Such paradigms produce stereotypical views of members of

*Gavin Steingo*

Third World cultures, who are always seen as representatives driven solely by the cause of vindicating their own cultures" (138). Chow argues that the "paradigm of collectivity" in China cannot be understood outside of the history of Western imperialism: emerging from imperialism, she argues, China had no alternative but to "'go collective' and produce a 'national culture'" (138). Hence, if Walkman listening in China is essentially a statement that "I am not there, not where you collect me" (145), this implies a completely different scenario from the "alienated" or "atomized" individuals described by writers commenting on the West.

I am also not the first to present an alternative theorization of automobile listening. In an excellent article about sound in the greater Johannesburg metropolitan area (including Soweto), Xavier Livermon (2008: 275) challenges Bull's description of cars as mobile bubbles and argues that in South Africa, the car "becomes a space of high sociality, with interactions between members of the vehicle and those who are outside the vehicle." As he observes, car horns, music, and human voices are ejected outward from the car and toward the public. My own fieldwork largely corroborates Livermon's observations. But I have also argued elsewhere—based on the differences in automobile habitation in Johannesburg's townships and suburbs—that in South Africa motor vehicles encapsulate the dialectic of private and public, of safety and danger (Steingo 2016). In other words, automobile use in South Africa is variously antisocial and social; the distinction between social and antisocial is, furthermore, often blurred.

Accounts of media use in Africa also paint a very different picture of claims surrounding music's unlimited availability and ubiquity. In 2012, Adam Haupt observed that thirteen million South Africans were using the Internet on mobile phones for instant message services such as Mxit. But despite an apparent narrowing of the so-called digital divide between global North and South, "mobile access to the internet has limitations, depending on the level of sophistications of one's mobile phone" (Haupt 2012: 115). In general, text-based applications are far easier and cheaper to perform than more complex tasks such as downloading or uploading music or video clips (Haupt 2012: 115). The point, then, is not so much about technology *lack* as about differential uses of globally circulating technologies.

At least for the people I usually interact with in Soweto, music is almost always shared off-line through storage devices (see Steingo 2015). The prevalence of offline exchange has been documented by other researchers in South Africa (Schoon 2011), and Christopher Kirkley observes a similar phenomenon in northern Mali, where people use their cell phones "as portable hard drives" and exchange music primarily through Bluetooth, which does not require Internet or phone service (see Rohr 2013).

Tom Astley's meticulous research on music circulation in contemporary Cuba offers yet another relevant case in point. "In Cuba," Astley observes, "the internet is far from a ubiquitous cultural source, and so conceptions of social networks and immediate access to cultural texts that make sense of online file-sharing are not applicable in quite the same way" (Astley 2016: 16). Astley notes that file sharing in Cuba typically takes place off-line; information "is regularly swapped [via USB devices] through meetings in private spaces." While the ubiquity of off-line networks is partly attributable to government censorship of the Internet, Astley makes the interesting observation that peer-to-peer sites are in fact "blocked primarily because of their *use of bandwidth*" (Astley 2016: 24, emphasis added).

As these case studies from South Africa, Mali, and Cuba illustrate, music is not necessarily always already there. Although the MP3 has come to represent the apotheosis of fluidity and frictionless flow—the long dreamed-of cloud or the celestial jukebox—it is precisely the MP3 format that now "circulates" *off-line* in various communities; that moves awkwardly, slowly, in fits and starts.[11] If we are willing to acknowledge that the situation in South Africa is a *product* of global modernity rather than merely a technological lag, then once again we are launched back into the profound theoretical discussion of how best to characterize modernity in the first place.

On the topic of hearing loss as it relates to efficiency and optimization, a global South perspective would necessarily have to take into consideration the long-standing theoretical debate on biopower. I have already suggested that the key tenet of modern biopower—that is, to "'make' live and 'let' die"—is not easily applicable to many parts of the world, especially not to my primary field site of South Africa. This is not surprising, especially if one considers that when Foucault developed his famous theorization of biopolitics and the racial state in the 1970s, he "had not a word for South Africa, the era's only example of 'actually existing' legal segregation" (Mbembe 2011: 89).

In "The Miner's Ear," Morris notes that in apartheid-era South Africa the procedures associated with Foucault's notion of biopower were "implemented only partially, and primarily for white populations. Black miners were treated as expendable and received relatively little biopolitical investment. They worked until illness made them, from the mine company's perspective, useless, at which point the miners were abandoned to fate" (Morris 2008: 110–11). Hence, although the loud sounds in gold mines often caused severe hearing loss, and although this hearing loss "increase[d] the miner's

*Gavin Steingo*

vulnerability to yet more injury," very few steps were taken to protect the miner's hearing or even his life. In contrast to the biopower theorized by Foucault, the apartheid state acted according to a logic of *superfluity* in which black life was considered excessive and "constituted wealth that could be lavishly spent" (Mbembe 2008: 43).

## SOUND STUDIES REMAPPED

This chapter has focused on the contrast between sound as it is theorized within the nascent discipline of sound studies and the ways that sound is experienced and practiced in various African locales. On the one hand, "sound students" (as Jonathan Sterne [2012b: 3] suggests calling scholars of sound studies) can only theorize sound in the way that they do because they ignore its differential use in Africa and other parts of the world. This implies an *empirical* shortsightedness in which non-Western sound practices are ignored, bracketed, or occasionally mentioned only as brief caveats (as in the cases of Kassabian, Gopinath, and Stanyek) before returning to business as usual.

On the other hand, it may be possible to attribute the shortsightedness of recent studies to long-standing metaphysical assumptions about the nature of sound. For example, the notion of automobiles as "cocooned" acoustic environments seems to realize—in material form—a particular metaphysical characteristic of sound, exemplified by a passage from Marshall McLuhan and Quentin Fiore's famous book *The Medium Is the Message*: "We are enveloped by sound, it forms a seamless web around us. We can't shut out sound automatically. We simply are not equipped with earlids" (McLuhan and Fiore 1989: 111). The first two items in Sterne's oft-cited "audiovisual litany" closely resemble the notion of sound as an envelope: (1) "Hearing is spherical, vision is directional"; and (2) "Hearing immerses its subject, vision offers a perspective" (Sterne 2003: 15). As Sterne argues, such assumptions about hearing and seeing are by no means obvious or scientifically grounded. Instead, he says, "The audiovisual litany is ideological in the oldest sense of the word: it is derived from religious dogma" (16).[12]

In a similar manner, the "observation" that sound has become infinitely available and ubiquitous may be predicated on nonempirical assumptions about the metaphysics of sound itself. Amy Cimini and Jairo Moreno (2016: 37) have recently argued that within the Western tradition sound is often understood under the sign of the "fiduciary"—that is, as a conjunction of fidelity and faith. They suggest that Western thinking about sound is largely governed by this fiduciary logic, according to which "perception always already believes that *it* senses and *in* what it senses." From phenomenology and performance theory to the affective turn, the listener is asked to trust *what* she hears and, before that, *that* she hears. In short, the assumption is that there will be sound.

The Northern bias of sound studies that this book seeks to overcome is not, then, limited to empirical data. It extends to a host of metaphysical assumptions about what and how sound is.

A more comprehensive study than this one would need to delve more deeply into the ways that sound and listening are conceptualized by a wide range of people on the African continent. It would also require a careful ontological investigation of what listening is and how sound is constituted across subjects, technologies, and platforms. Without exaggerating the extent to which language determines auditory experience, it is nonetheless worth pointing out that in Zulu the term *ukuzwa* can refer to both hearing through the ears and any tactile sensation—what in English we would call "feeling." In this sense, the insight that sound is not exclusively an auditory relation—that, for example, deaf people can experience sound through tactile vibrations of the skin (cf. Friedner and Helmreich 2012)—is quite literally registered by Zulu speakers in a single word. But to complicate matters further, the root *-zwa* can also mean "understand."[13] Hence, the word can refer to something more than a "brute" physical sensation. A second aspect worthy of comment (and, I believe, further study) is that in Zulu cosmology, sound does not affect people exclusively through air-mediated vibrations. Instead, specific sounds are thought to have qualities that act on people mysteriously and at a distance. Many of my interlocutors in Soweto understand sonic causation in terms of spiritual forces and not mere mechanical action. Examples of nonphysical causation are everywhere in anthropological studies of southern Africa (see, e.g., Berglund [1976] 1989; Hammond-Tooke 1981).

But a full account of how sound is conceptualized in Africa lies far beyond the scope of this chapter. In closing, I instead attempt something more modest and briefly discuss a final possible reason why Africa has been largely omitted from the sound studies canon. While the lacuna may simply belie a troubling indifference toward the world's second-largest continent, there may be a more theoretically significant reason for Africa's absence. As Ferguson has observed, and as I have illustrated in this chapter, Africa is a particularly inconvenient example for many globalization narratives. Most "ostensibly all-encompassing narratives," writes Ferguson (2006: 25), "manage to char-

*Gavin Steingo*

acterize 'the globe' and 'the entire world' in ways that say almost nothing about a continent of some 800 million people that takes up fully 20 percent of the planet's land mass." And they do this, Ferguson argues, because Africa does not fit into the narrative of global *convergence*. While Ferguson focuses on narratives of capital flows and governance, his observations can easily be applied to sound and listening. When it comes to mobile and private listening, the ubiquity and availability of music, or a biopolitical investment in the human sensorium, Africa is an inconvenient example that shows no signs of converging with Euro-America.

The future, of course, is uncertain. With structural adjustment policies previously targeted at the global South now returning with a vengeance to the global North, the idealized listener of sound studies is quickly becoming antiquated, even farcical. If there is any convergence in the future, it seems more likely that it will entail Euro-America's movement in the direction of Africa rather than the other way around. Perhaps Rem Koolhaas is on to something when he resists the idea that major African cities are simply "en route to becoming modern. Or, in the more politically correct idiom, that they are becoming modern through a valid, African way" (Koolhaas n.d.: 138, quoted in Mbembe and Nuttall 2008: 4). Perhaps he is right to speculate that African cities "represent a crystallized, extreme, paradigmatic set of case studies of cities at the forefront of globalizing modernity"—a modernity that is quite at odds with the horizon assumed by most convergence narratives. Perhaps, finally, Koolhaas will be proved correct when he suggests that a city such as Lagos "is not catching up with us" but, instead, that "we may be catching up with Lagos" (Koolhaas n.d.: 85, quoted in Mbembe and Nuttall 2008: 4). Only time will tell. But for now, it is certainly possible to observe a very specific set of sonic practices in Africa—practices that cannot be reduced to those in the global North.

The title of this chapter is a reference to Veit Erlmann's magisterial *Reason and Resonance* (2010), which investigates the underestimated role of sound and hearing in the construction of Western knowledge—that is to say, the intertwining of reason (knowledge) with resonance (sound). Or, in Sterne's (2003: 2) words, "As there was an Enlightenment, so too was there an 'Ensoniment.'" This chapter has argued that there are other resonances, as well—ways to understand sound and listen that cannot be reduced to Western practices or Enlightenment history. These other resonances may be largely absent from recent literature in sound studies, but they are everywhere audible in Africa and, perhaps, in most of the world.

## Notes

1. He describes this final stage as a "technological regression" or "devolution," since "the Walkman is a cassette recorder *minus* the recording function and the speaker" (Hosakawa 1984: 168).

2. Bull (2001: 358–59) quotes Jacobson (1999): "In the car you are physically co-cooned. . . . It is the last private space in an overwhelmingly public world, the nearest we get to a lavatory at the bottom of the garden, where people once went to have a little time to them." Bull provides 2000 as the date of Jacobson's piece but does not include the piece in his reference list. Fortunately, I able to find the passage in question online.

3. For a different take on sound and cars by LaBelle, see LaBelle 2008.

4. Burkhart and McCourt (2006: 4) note, "The Recording Industry Association of America (RIAA), the powerful music-industry trade group and principal agent in the culture industry's technology wars, used the Celestial Jukebox metaphor in 1989 and 1990 in a rhetorical effort to gain legislative support for digital-performance rights on the Internet."

5. This actually is a side point for Morris. His main argument is that, in the case of streaming music platforms, listeners have a different relationship to music because they do not "own" it. In a later text, Morris in fact cautions us that "the digital music commodity is not as intangible as it is sometimes presented. . . . Information is not immaterial or dematerialized. . . . It is always embodied or expressed materially (Wark 2006: 173). Rather than dematerialization, digital music is a *rematerialized* commodity, one whose materials bring new sources of value for listeners, companies, and music itself" (Morris 2015: 14). Even so, Morris maintains that "digital music is also a highly *fluid*, *mobile*, and *social* commodity" (12).

6. This colleague, Kassabian tells us in a footnote, was the (Italian) academic Massimo De Angelis (see Kassabian 2013: 119n1).

7. Although Africa is absent from their introduction (as well as from their two-volume *Handbook*), it is worthwhile mentioning a curious endnote that closely resembles Kassabian's rhetorical gesture. Gopinath and Stanyek (2014: 30n11) write, "Such an understanding of the notion and experience of portability certainly does not encompass the entirety of mobile music's history, which can be said to take place at different times and in different ways all over the world. Indeed, only recently have locations in developing countries become unprecedentedly suffused with portable sound media (on account of the mobile telephone). Nonetheless, this critical aspect of mobile music's development has ramifications beyond the specific discursive regional problematic under consideration in this chapter." Like Kassabian, Gopinath and Stanyek admit possible differences in other parts of the world ("the notion and experience of portability certainly does not encompass the entirety of mobile music's history"), imply a trajectory of global convergence ("developing countries [have recently] become unprecedentedly suffused with portable sound media"), and then

*Gavin Steingo*

essentially dispense with global difference as not fundamental to their primary argument ("*Nonetheless,* this critical aspect of mobile music's development has ramifications beyond the specific discursive regional problematic under consideration in this chapter"). But here again the authors too quickly pass over the global South. Is it adequate to mention global difference as mere dispensable caveat? I think it unwise to believe so.

8. As Mills (2011: 24) notes, Michael Riordan and Lillian Hoddeson (1998: 205) refer to the "'relentless progress of miniaturization' following the 1948 invention of the transistor."

9. Of course, drivers are warned, as well. For more on walking and pedestrianism in Soweto, see Steingo 2016: 206–11.

10. The title of Bickford's essay is: "Earbuds Are Good for Sharing: Children's Headphones as Social Media at a Vermont School."

11. I thank Mary Caton Lingold for discussing this point with me. Mary Caton's use of the word "apotheosis" helped crystalize the idea toward which I had been working. Sterne (2012a) has shown in the meticulous detail that the *invention* of the MP3 was closely connected to online circulation.

12. Sterne (2003: 16) continues that the audiovisual litany "is essentially a restatement of the longstanding spirit/letter distinction in Christian spiritualism."

13. It is nonetheless interesting to note that Zulu does draw a distinction that is very similar to the oft-noted "hearing" versus "listening." In Zulu, *zwa* generally means "hear," while *lalela* means "listen."

## References

Adorno, Theodor W. [1927] 2002. "Curves of the Needle." In *Essays on Music,* ed. Richard Leppert, 271–76. Berkeley: University of California Press.

Astley, Tom. 2016. "The People's Mixtape: Peer-to-Peer File Sharing without the Internet in Contemporary Cuba." In *Networked Music Cultures: Contemporary Approaches, Emerging Issues,* ed. Raphaël Nowak and Andrew Whelan, 13–30. London: Palgrave Macmillan.

Baudrillard, Jean. 1993. *Symbolic Exchange and Death.* London: Sage.

Beer, David. 2007. "Tune Out: Music, Soundscapes, and the Urban *Mis-en-Scène.*" *Information, Communication, and Society* 10, no. 6: 846–66.

Berglund, Axel-Iver. [1976] 1989. *Zulu Thought Patterns and Symbolism.* Bloomington: Indiana University Press.

Bickford, Tyler. 2014. "Earbuds Are Good for Sharing: Children's Headphones as Social Media at a Vermont School." In *The Oxford Handbook of Mobile Music Studies,* vol. 1, ed. Sumanth Gopinath and Jason Stanyek, 335–55. Oxford: Oxford University Press.

Bijsterveld, Karin. 2006. "Listening to Machines: Industrial Noise, Hearing Loss and the Cultural Meaning of Sound." *Interdisciplinary Science Reviews* 31, no. 4: 323–37.

Bijsterveld, Karin. 2010. "Acoustic Cocooning: How the Car Became a Place to Unwind." *Senses and Society* 5, no. 2: 189–211.

Bijsterveld, Karin, Eefje Cleophas, Stefan Krebs, and Gijs Mom. 2014. *Sound and Safe: A History of Listening Behind the Wheel*. Oxford: Oxford University Press.

Bull, Michael. 2001. "Soundscapes of the Car: A Critical Study of Automobile Habitation." In *Car Cultures*, ed. Daniel Miller, 185–202. New York: Berg.

Bull, Michael. 2004. "Automobility and the Power of Sound." *Culture, Theory, and Society* 21, nos. 4–5: 243–59.

Bull, Michael. 2005. "No Dead Air! The iPod and the Culture of Mobile Listening." *Leisure Studies* 24, no. 4: 343–55.

Bull, Michael. 2007. *Sound Moves: iPod Culture and Urban Experience*. New York: Routledge.

Burkart, Patrick, and Tom McCourt. 2006. *Digital Music Wars: Ownership and Control of the Celestial Jukebox*. Lanham, MD: Rowman and Littlefield.

Chambers, Iain. 1990. "A Miniature History of the Sony Walkman." *New Formations* 11 (Summer): 1–4.

Chow, Rey. 1990–91. "Listening Otherwise, Music Miniaturized: A Different Type of Questions about Revolution." *Discourse* 13, no. 1: 129–48.

Cimini, Amy, and Jairo Moreno. 2016. "Inexhaustible Sound and Fiduciary Aurality." *boundary 2* 43, no. 1: 5–41.

du Gay, Paul, Stuart Hall, Linda Janes, Hugh Mackay, and Keith Negus. 1997. *Doing Cultural Studies: The Story of the Sony Walkman*. London: Sage.

Erlmann, Veit. 2010. *Reason and Resonance: A History of Modern Aurality*. New York: Zone.

Ferguson, James. 2006. *Global Shadows: Africa in the Neoliberal World Order*. Durham, NC: Duke University Press.

Foucault, Michel. 2003. *Society Must Be Defended: Lectures at the Collège de France, 1975–76*, ed. Mauro Bertani and Allesandro Fontana, trans. David Macey. New York: Picador.

Friedner, Michele, and Stefan Helmreich. 2012. "Sound Studies Meets Deaf Studies." *Senses and Society* 7, no. 1: 72–86.

Gopinath, Sumanth, and Jason Stanyek, eds. 2014. *The Oxford Handbook of Mobile Music Studies*, vol. 1. Oxford: Oxford University Press.

Hammond-Tooke, W. D. 1981. *Boundaries and Belief: The Structure of a Sotho Worldview*. Johannesburg: Witwatersrand University Press.

Haupt, Adam. 2012. *Static: Race and Representation in Post-Apartheid Music, Media and Film*. Cape Town: HSRC Press.

Holmquist, Lars Erik. 2005. "Ubiquitous Music." *Interactions* 12, no. 4: 71–78.

Hosakawa, Shuhei. 1984. "The Walkman Effect." *Popular Music* 4: 165–80.

Jacobson, Howard. 1999. "Road Rage." *The Independent*, December 7. Accessed June 18, 2018. http://www.independent.co.uk/news/people/profiles/road-rage-743173.html.

Kassabian, Anahid. 2013. *Ubiquitous Listening: Affect, Attention, and Distributed Subjectivity*. Berkeley: University of California Press.

Koolhaas, Rem. n.d. "Lagos: How It Works." Unpublished ms., copy in the personal collection of Achille Mbembe.

*Gavin Steingo*

Kracauer, Siegfried. [1927] 1995. *The Mass Ornament: Weimar Essays*, ed. and trans. Thomas Y. Levin. Cambridge, MA: Harvard University Press.

Livermon, Xavier. 2008. "Sounds in the City." In *Johannesburg: The Elusive Metropolis*, ed. Sarah Nuttall and Achille Mbembe, 271–84. Durham, NC: Duke University Press.

LaBelle, Brandon. 2008. "Pump Up the Bass—Rhythm, Cars, and Auditory Scaffolding." *Senses and Society* 3, no. 2: 187–204.

LaBelle, Brandon. 2010. *Acoustic Territories: Sound Culture and Everyday Life*. London: Continuum.

Mbembe, Achille. 2008. "Aesthetics of Superfluity." In *Johannesburg: The Elusive Metropolis*, ed. Sarah Nuttall and Achille Mbembe, 37–67. Durham, NC: Duke University Press.

Mbembe, Achille. 2011. "Provincializing France?" *Public Culture* 23, no. 1: 85–119.

Mbembe, Achille, and Sarah Nuttall. 2008. "Introduction: Afropolis." In *Johannesburg: The Elusive Metropolis*, ed. Sarah Nuttall and Achille Mbembe, 1–33. Durham, NC: Duke University Press.

McLuhan, Marshall, and Quentin Fiore. 1989. *The Medium Is the Message*. London: Touchstone.

Mills, Mara. 2011. "Hearing Aids and the History of Electronics Minimization." *IEEE Annals of the History of Computing* 33, no. 2: 24–44.

Morris, Jeremy. 2011. "Sounds in the Cloud." *First Monday* 16, no. 5. Accessed June 18, 2018. http://firstmonday.org/article/view/3391/2917.

Morris, Jeremy. 2015. *Selling Digital Music, Formatting Culture*. Berkeley: University of California Press.

Morris, Rosalind. 2008. "The Miner's Ear." *Transition* 98: 96–115.

Polgreen, Lydia. 2015. "A Music-Sharing Network for the Unconnected." *New York Times*. June 4. Accessed June 18, 2018. http://www.nytimes.com/2015/06/07/magazine/a-music-sharing-network-for-the-unconnected.html.

Riordan, Michael, and Lillian Hoddeson. 1998. *Crystal Fire: The Invention of the Transistor and the Birth of the Information Age*. New York: W. W. Norton.

Rohr, Rachel. 2013. "Mali Music Culture Defined by the Cellphone." *Here and Now*, February 25. Accessed April 25, 2015. http://hereandnow.wbur.org/2013/02/25/mali-cellphone-music.

Schoon, Alette Jean. 2011. "Raw Phones: The Domestication of Mobile Phones amongst Young Adults in Hooggenoeg, Grahamstown." MA thesis, Rhodes University, Grahamstown, South Africa.

Simmel, Georg. [1903] 1997. "The Metropolis and Mental Life." In *Simmel on Culture: Selected Writings*, ed. David Frisby and Mike Featherstone. London: Sage.

Steingo, Gavin. 2015. "Sound and Circulation: Immobility and Obduracy in South African Electronic Music." *Ethnomusicology Forum* 24, no. 1: 102–23.

Steingo, Gavin. 2016. *Kwaito's Promise: Music and the Aesthetics of Freedom in South Africa*. Chicago: University of Chicago Press.

Sterne, Jonathan. 2003. *The Audible Past: Cultural Origins of Sound Reproduction*. Durham, NC: Duke University Press.

Sterne, Jonathan. 2012a. *MP3: The Meaning of a Format*. Durham, NC: Duke University Press.

Sterne, Jonathan. 2012b. "Sonic Imaginations." In *The Sound Studies Reader*, ed. Jonathan Sterne, 1–17. New York: Routledge.

Van Buskirk, Eliot. 2006. "Celestial Jukebox Falls to Earth." *Wired*, June 12. Accessed June 18, 2018. http://www.wired.com/2006/06/celestial-jukebox-falls-to-earth.

Virilio, Paul. 1989. "The Last Vehicle." In *Looking Back on the End of the World*, ed. Dietmar Kamper and Christoph Wulf, 106–19. New York: Semiotext(e).

Wark, McKenzie. 2006. "Information Wants to Be Free (But Is Everywhere in Chains)." *Cultural Studies* 20, nos. 2–3: 165–83.

Weinberg, Gil. 2003. "Interconnected Musical Networks: Bringing Expression and Thoughtfulness to Collaborative Group Playing." PhD diss., MIT, Cambridge, MA.

Williams, Raymond. 1974. *Television: Technology and Cultural Form*. Hanover, NH: Wesleyan University Press.

Wolk, Douglas. 2009. "The Future of Music: The Celestial Jukebox." *Upstart Business Journal*, March 30. Accessed June 10, 2017. http://upstart.bizjournals.com /companies/innovation/2009/03/30/The-Celestial-Jukebox.html?page=all.

*Gavin Steingo*

# 2

# Ululation

*Louise Meintjes*

.....

For migrant Zulu men at home in Msinga, in rural KwaZulu-Natal, *ngoma* performance is a form of participatory politics with regard to community life as it offers a way of being in the world. Imagine the politics: Mboneni, curtailed in a moment of improvised dare; Uzowotha dancing, and nothing is spoken; Zabiwe competitively slicing through Siyazi's dance that day; Mdo strutting, calling out praises, struggling this year against the virus; spindly Sono taking the place of his assassinated father in the dance. Notice ways of being and the pleasures being had: boys parroting the adult dancers, and ululating mothers exhorting their sons. Girls watching. A dancer boasting. An admirer deputizing her sister to present gifts to a dancer: matches, cigarettes, a T-shirt. Zabiwe singing with the eloquence of men. Ntibane trumping the moves of his friend. The elders advising and blessing. The granny crisscrossing the dance floor, crisscrossing, crisscrossing, as the high summer afternoon wind blows.

Zulu men's song and dance called ngoma grew out of South Africa's migrant labor system through the twentieth century. It is a recreational, high-prestige form organized around competitive all-male teams who dance for their communities at homecoming times and for their fellow migrants in Johannesburg. The competition within and among teams is coupled with intense camaraderie,[1] ngoma being a practice that cultivates masculine ideals.

In a Zulu ngoma aesthetic of "no gaps"—a dense layering of synchronically overlapping performances—the crisscrossing granny follows her own trajectory through the outdoor performance space ringed by the team's fans,

FIG 2.1. Ma Soshangani Zulu, Keates Drift, Msinga, December 25, 2008.
Photograph by the author.

friends, and family. As she hobble dances with her walking stick, weaving around the dancers in the middle of the arena, she ululates. She continues all afternoon. "Zizekubani? Ezakabani?" (Who are the girls coming to see/ visit? Where are they coming from?) she inserts rhythmically between her trills and over the men's singing as she passes close to the dancers.[2] She plays with phrases conventionally asked of girls coming to represent their courting sisters at a boys' homesteads. Two dancers enter her rhythmic verbal play, exchanging stock ritual phrases with her. It is a playful passing duel to which only those in proximity are privy. Later, the dance captain spots a girl checking out the dancers as he leads the team in song. "Girl, choose your boyfriend, please. The room is full of boys. Girl, where is your boyfriend?" he inserts into the stream of his narrative, improvising in song.

Ululation is a high-pitched trilling by means of oscillation of the tongue. The crisscrossing granny's ululations, along with those of other women in the "audience," contribute formal components to ngoma's acoustic performance. They fill in temporal gaps; they fill out the sound spectrum in a band-

*Louise Meintjes*

width cooccupied by men's exclamatory dental whistles and the dance team captain's metal whistle ushering commands. They excite the texture with the attack and vibrating quality of their melodic lines. The timing of their outbursts often delineate sections in men's performance. Ululation builds a relationship to the performance to which it responds and with which it overlaps. Perhaps it is something akin to signifyin(g).[3]

I am struck by the near-omission of ululation from ethnomusicological scholarship, as though ululation were postperformance clapping. Where it appears, it does so cursorily as extraneous, if notable, noise.[4] I argue that it is an artful presence, an unmarked gendered feature of ngoma that in effect makes the form.

My metaproject here is to notice the way in which the global South's relationship to sound studies remains largely unmarked, as though the global South were sound studies' postperformance clapping. Via ethnographic close analysis of ululating Zulu women in rural KwaZulu-Natal, I hope to contribute toward getting the global South more widely heard as a cocreator of sound studies from a place of struggle. Marking the global South makes explicit the requirement that sound studies be gendered; its discourse racialized; its relationality recognized; and its sounds heard as particular.

Ululation is a performance in itself. It projects the presence of women through the rasp of the high summer afternoon wind, across wide outdoor arenas, over the booming drums and the singing and exclamations of the dancers. It does so by its characteristic high pitch, loud volume, timbral intensity, and sustain (Jacobs 2008). There are intermittent ululators, as there are virtuosos, such as Ma Soshangani Zulu, crisscrossing as the wind blows. There are liquid warbles, wavering oscillations, and stuttering trills; cascading melodic lines, pithy rhythmic iterations; shrill whistles and reverberating upper mid-range calls; piercing interjections and soaring projections. Some have deep lungs; others, rapid-fire tonguing technique.

One ululates for or on behalf of: *ukikizela*. The verb *ukukikiza* appears commonly in everyday talk with the applicative suffix *-ela*, which identifies the ululator as acting in relationship to someone or something.[5] Two chatting women turn to watch the performers. They respond to the lead singer's phrase, partially covering their mouths decorously when they ululate. The first is the singer's mother ululating for her son; the second is his sister.

Ululators time their utterances in relation to the singer-dancers' performance. The mother times her ululation to cascade over the final phrase of the singer-dancers' chorus, initiating it precisely in a sub-phrase gap. She utters three phrases, consistent in pitch, timing, and melodic contour.

The team captain's whistle shares the bandwidth, further exciting the sound.

With staggered onsets, each woman's ululation is synchronized with other ululations in a polyphonic overlap. The singer's sister begins her phrase as her mother's first utterance completes its descending tail. Her pitch and melodic contour is different from her mother's, and she follows her ululation with a texted exclamation. In turn, her mother punctuates the call with agreement, musically timed. Here we hear the women relating acoustically to the lead singer, as well as to each other.

The timing of specific ululations can mark and cultivate relationships among women while they build ngoma's form. Similarly, while ululations are not coordinated across the whole gathering, one set of ululations can provoke another somewhere else in the crowd, producing a kind of collective effervescence.

In the process of ululating, women are exhorting, competing, affirming, agreeing, celebrating, proclaiming, encouraging, anticipating, or praising, sharing in a mature women's world while marking their relationships to one another and to men. Danced eruptions likewise mark individual women's relationships to specific men. Khulukuthu, the wild one, executes a magnificently fiery sequence. His wife, Makahlela, bursts out of the crowd and rushes into the arena swinging her handbag like a lasso. She veers toward him, loops back to the encircling crowd, dances an exuberant step or two, then merges back into the audience. I think of women's spurts out of the crowd into the arena as danced ululations—the voice, in other words, as a body-voice.

Ululation's defining technique is the oscillation of the tongue, using vocables rather than sung text. In ululation as a tongued body-voice performance, the sound producing technology is artful. The tongued body-voice is also the resonator, amplifier, and medium of distribution. In other words, the tongued body-voice is technology, technique, and techne (craft).[6] As sound studies has canonized into excellent readers, special issues, and handbooks, it has predominantly attended to sound technology (and sometimes leaves out the sound). Gavin Steingo and Jim Sykes remark in this volume that, written from and mostly about the North, sound studies erroneously presumes ubiquity, newness, efficiency, and abundance of technology—digital technology—irrespective of place. How might we remap the sound studies terrain not only to better include the sonic practices of the South, but also to rework the terms into ones that do not position the global South abjectly?[7] Can one hear the global South ululating?

*Louise Meintjes*

The tongued body-voice as technology embeds unique affective biographies and political histories in each utterance. It is always already gendered, raced, and implicated in struggle. The crisscrossing granny lives in Msinga, an area that was designated an African reserve as early as 1848. Colonials pushed Africans off fertile land into this rugged area, which has been maintained as a "reservation" through various dispensations ever since. Today the area holds at least twice the number of people and cattle than the land can sustain (Association for Rural Advancement [AFRA] newsletter no. 14, in "All CAP/ Mdukatshani Newsletters from the 1960s till the Late 1990s" 2016). Households rely on multiple sources of income: small-scale farming (including livestock) for consumption and sale; earnings from agricultural labor nearby or from migrant labor in the cities and industrial areas; and any other small or temporary options that present themselves, such as driving taxis. Illegal activities such as marijuana growing and gun running provide income for some. Commerce is sparse. For many families, welfare and social support grants from the state are essential.[8] Msinga is the fourth poorest out of 227 local municipalities in South Africa, with 86 percent of the population living below a lower-bound poverty line (Noble et al. 2014).[9] Unemployment lies at 69 percent in the municipality that this ngoma community calls home.[10] Households struggle against daunting odds. The women live polygamously (mostly, and especially mature women). They live under the jurisdiction of a male chiefdom and the state, in a community organized through a system of headmen. In an area impoverished by a history of colonialism and apartheid, and their violent legacy (including the HIV epidemic), women bear the weight of maintaining homesteads in the context of largely male migrant labor. Young women who seek work elsewhere often leave primary childcare to grandmothers. Women ululate out from this history, carried in the body-voice.

Specific ululations also voice specific biographies and specific violent politics. It is December 25, 2002, and the team has entered the arena to begin its display. The team members are seated in two long lines, ready to sing. First, Vice-Captain Mdo—the singer whose mother and sister ululated for him—inspects his dancers' ranks, strutting up and down between the lines, admonishing slouchers. Next he sings the team into song, pacing in front of them. He walks to the edge of the arena, to where the crowd gathers to watch. There he stands erect atop a rock, surveying his dancers from afar. "Moliva!" he calls out to them, incanting the name of their place. "Shiya!" they respond, saluting him from the center of the dusty arena. He paces farther out of the arena. Where is he going? (As a lead singer, he always struts, commandeering the dance arena by pacing while singing instructions to the team. But he has never walked this far away.) He saunters back to the rock. Unusually, he is stretching out the pauses. He sings out, lines rich in masculinist metaphor and innuendo, drawing on the poetic reservoir of home and displaying his verbal agility. "Yeyi, black bull, come and surprise me!" he sings, as if provoking a good dancer into a challenge. He stalls for a moment, surveying the scene. "Thank you, my children [dancers]. Respect yourselves," he shouts. "Yes!" he exclaims, before rattling off stock phrases that reference the district: "I tell you, if you respect yourself, even your dancing will respect you and your dance will get respect. Never, never, never! [i.e., Don't make any mistakes]," he incants, although only those in his proximate vicinity hear him. Ululations fill the space like sonic filigree. "Hey, father!" calls a woman from his family, addressing him with respect. Soon he resumes his strutting, following the arena's circumference, outlined by the watching crowd, while women ululate and he sings, and the dancers answer him from the center. Mdo works against the acoustic constraints of the outdoor event to render his signature singing, risking his power, playing with his control at the edge. He pushes his voice to its acoustic limits. Exacerbating the acoustic difficulty, he paces to the limits of the arena. Mdo takes his own style to the extreme.

Did I notice, a dancer said some years later, that Mdo did not dance that day in 2002? He did not insert a single virtuosic solo dance, as is usual for leaders. When dancing on that day exceeded Mdo's capacity, he instead focused his authority in his voice. Mdo asserted undisputable leadership by means of a vocal performance made dramatic by his use of space and his poetic language, even with a weakened voice. He used the resonance of open vowels to throw his voice, although he could not fully sustain his tone and pitch. He switched to the rasp of his throat for contrast. He shifted the form, adding variation and interest by inserting incanted praise from his distant rock. ("Never, never, never, people of esiPongweni!") He made his sitting dancers wait, just too long, until they called on him to let them begin, texturing the event with their interjections. He thereby commanded the team and directed the flow of the event. In turn, the team honored his new limitations by refraining from exhorting him to dance yet admiring him with shout-outs of his dancing praise name while he strutted around their cluster. Nothing was spoken. Mdo's vulnerable condition—he was struggling against the virus—was publicly articulated and his reduced capacity managed while the team members sustained their ordinary social relationships, including with their friend. Throughout, he elicited a plethora of ululation, whether uttered in appreciation of the drama he crafted, in knowing support of the effort this took him, or to cover for the signs of his stigmatizing struggle.[11]

Ululation has the capacity to cut through, to enact a genre appropriately, to fill in and flow over. Its passing specificity is there to be heard under the cover of ambiguity and while making a musical contribution. Fans appreciate the artistry of ululation. Ma Soshangani Zulu, the crisscrossing granny, has told me about her unmatched vocal renown in the valley and her skill as a dancer during her girlhood. Ngoma performers also consider ululation a component of their art. When the Umzansi Zulu Dancers, the professional subgroup of the community team, self-produces their CDs, ululation contributes an aesthetic effect. In "Zindala Zombili," for example, the group plays with various quivering textures and cascading contours in the upper bandwidth of their sound, overdubbing a range of blown whistle sounds and ululations.[12] Similar to the valued stridence of a Zulu "traditional" guitar sound that enables the instrument to cut through a busy mix, whistles and ululations in the high register cut through the sound and excite it. The way ululation is integrated highlights its aesthetic coherence with Zulu musical style. It is also used as a compositional element: overdubbed ululations are positioned to mark the verse-chorus form in "Khuzani."[13] Here, the improvised timing of women's response, such as of the mother initiating her ululation in the chorus's sub-phrase gap—is reproduced as a regularized framing device. One also gets a sense of how present ululation is in the sound for ngoma dancers. To the ululations on "Khuzani" the group added reverb and echoes that one would never hear in an open acoustic environment of ululation performance (Doyle 2005). By means of this reverb, the Umzansi Zulu Dancers are enhancing a valued quality of ululation, namely its sense of projection, and pitching the idea of outdoor rural sociality. The inclusion of ululation on these recordings, as well as the way in which it is produced, points to the significant presence of women in the aesthetics—and the local politics—of ngoma events.

## ULULATING AS AND ON BEHALF OF THE GLOBAL SOUTH

Billions of women in the global South ululate out into the world from a range of religious, class, ethnic, national, historical, and gendered positions. Techniques vary, whether executed with the aid of jaw movement, as for some Zulu women, or not. Joel Kuipers (1999) describes the Indonesians flicking the tip of the tongue against the alveolar ridge of the palate. Jacobs (2008) finds Moroccans striking the teeth, alveolar ridge, or (with lateral oscillations) the corners of the lips with their tongues. Preferred vowels vary, and so, therefore, do lip positions, whether rounded (*o, u*) or widened (*ee, ay*). Melodic patterns vary, as does the textural density of polyphonic utterances. Generally, a ululation begins with a rapid onset—that is, a quick rise to the main sustained pitch—and tails off variously at the end of the breath. Whatever the variations in technique, ululation is an affirmative form of women's expression practiced across the Levant, the African continent, South and Southeast Asia, and Aboriginal Australia.[14] Weddings are the most widely reported venue, but others include parties, ritual, and performance events. Jacobs (2008) mentions ululation at a Palestinian funeral, but, as she notes, it was a funeral of a bride-to-be.

Ululation also erupts in political moments when women want to display communicative competence or get things done. Here ululation moves away from artistic performance, Mikhail Bakhtin's (1986) secondary genre, into the domain of everyday communication, Bakhtin's primary genre. I offer four examples.

At a time of domestic tension in the KwaZulu household that hosts me when I visit, I surreptitiously slipped the women extra cash. Rather than presenting it ceremoniously as a gift, I wanted to bypass the patriarch. But the two women had their own agendas. Rushing outside, they proclaimed news of our exchange across the valley in ululation. Minutes later, a neighbor arrived to hear the details, and soon the patriarch came scurrying home to investigate the to-do at his house.

Gavin Steingo witnessed academics ululating for their (female) dean at an event at the University of Pretoria. I would have expected the elite, cosmopolitan aspirations of female academics to preclude ululation. What I imagine was happening was that these female academics were publicly recognizing their dean respectfully as their titular head at the same time that they were celebrating that their dean was an African and a woman. Their gendered African voicing from the floor marked and collectivized the achievement on a campus that once denied them access.

To welcome Barack Obama onstage at the Global Entrepreneurship Summit in Nairobi in July 2015, the Kenyan television host and news anchor Julie Gichuru ululated for him.[15] Online, Kenyans debated the propriety of her performance voraciously, as they did, to a lesser extent, her skill. Critics' discomfort seemed to lie in the class disjuncture that Gichuru's performance represented (that ululating was "traditional" and she was British-educated); alternatively, celebrants tuned in to it as an Africanized voicing.

In the Algerian War of Independence (1954–62), ululation clearly had effects as it was banned by the occupying French (see Vergès 2015).

Despite the ubiquity of ululating, my search has rendered only two analyses focused on the technique and sound of ululation (Jacobs 2007, 2008; Kuipers 1999). Kuipers speculates on reasons that the significance of ululation as a vocal utterance has been missed: that it is uttered by women; it is neither music nor speech; and it is often not recognized as performance. Ululation, in other words, appears peripheral to genre constitution. There is, however, wide-ranging attention paid to keening and lamentation, which are predominantly associated with women; are neither music nor speech; and are not necessarily recognized as art. Perhaps lament positions women as victims, sufferers, and caretakers.[16] What's more, as an expression shared with women in the North, lament may have attracted more scholarly attention than the more affirmative voicing of women across the global South.

Analytical ears might scarcely be turned on to the acoustics of ululation, but there are many passing references to ululation in travelogues, missionary accounts, descriptions of African music and dance, and early ethnographies, as there are in studies of performance through the decades. Authors grant ululation a celebratory, noisy (wild), or resistant quality. Ululation also appears in metaphoric turns by poets, novelists, filmmakers, journalists, and theorists. The classic film *The Battle of Algiers* (1966) dramatizes collective resistance through the projection of voice. In the final scenes, a dark urban hillside dotted with domestic lights quivers acoustically. Unseen but ubiquitous women ululate. The effect is powerful for its all-encompassing resonant projection and for its untranslatability. In the context of the film, ululation is present as a warning, as an imminent threat of unshakeable resolve. It is the sound of a collective, impenetrable, unknowable (because inarticulate) but organized Other.

As in the film, critics and theorists invariably pick up on ululation's racialized politics, although they rarely differentiate ululated utterances or parse out their social value. For example, in her comparison of South African and southern U.S. literary texts, Sheila Smith McKoy coined the term "racial ululation," by which she means "a 'ritualized process of vocalizing a response to threats to white supremacist order' in order to 'maintain racial stratification and the disproportionate distribution of power that accompanies it'" (Smith McKoy 2001: 24–25, quoted in Steyn and Foster 2008). Melissa Steyn and Don Foster apply Smith McKoy by developing the idea of "white ululation" in South Africa, deploying but inverting its racialized profile. Looking at print journalism, they identify "white ululation" as a discursive repertoire among whites resistant to postapartheid transformation (Steyn and Foster 2008). These are not references to the sound of ululation nor to the pleasures of ululating. However one might critique this, or the theoretical value of the argument, or the extent to which the meaning of ululation needs to be broadened to work as a social metaphor here, Steyn and Foster usefully point to two key features of the practice. First, it is a racialized practice and situated in a world of radical power discrepancies. Second, reproducing a popular understanding of ululation, they presuppose that ululation is a way to talk back that is resistant. I am not sure it is resistant as much as it is a claim to ongoing participation.

What if we spin things around to treat ululation as an embodied practice that is crafted, subsumed through a lived history, enacted, and naturalized as affective, a tongued body-voice as technology? What if we approach it less as a framing device of men's articulate worlds, less as indexical of others' authority, than as a raced, gendered vocalizing presence saturated with struggle history?[17] Jacobs and Kuipers show the way. Kuipers (1999) approaches Indonesian ululation (Sumba island) not as an impoverished song genre but as audience response. Ululation represents gendered complementarity through interactive vocal overlap in ritualized speech situations (where men's performance is encouraged by women's ululation). He writes that by ululating women are "not primarily offering evaluations so much as expressions of their participation. In this sense, the ululators as 'audience'—actually an imprecise term—are more like coauthors engaged in a joint production and display of communicative competence" (Kuipers 1999: 492). Jacobs treats Levantine ululation as "sentiment-laden communication." For her, ululations are discourse markers and contextualization cues within an event. At the same time, they are construed meta-culturally as moral projections—that is, "Such performances are correct, appropriate, thoughtful, and good, hence motivating particular behaviors and desired social relationships within the unfolding circumstances" (Jacobs 2008: 178). Moroccan ululation (*zaghareet*) "expresses *farah*, 'joy' (to someone, about something)" (180).

Ululation amplifies and distributes a woman's voice. It is an insertion of praise while it also exhorts. It is an assertion of a relational presence—to and with other women as much as to performing men. A voice seemingly on the margins of (male-centered) performances, it is critical to proceedings and to the soundtrack of events, and it is heard within, over, and beyond an event. It is an announcement.

Ululation is an insertion made to be heard. That is why it needs to be artful. It asserts presence without saying something, referentially.[18] It is

*Louise Meintjes*

tongued body-voice. "Untexted" is not an absence, not a loss of voice; it is not an absence of proper language. To the contrary, it is the means of gaining recognition by means of being voice rather than logos. Sometimes it is music, not speech—the secondary genre of the stage; sometimes it is speech, not music—the everyday genre of a spontaneous outburst. It is, but it is not "art," while good ululators are appreciated. Ululations appear (being made sensible) on the stage, the street, the studio, and as everyday utterances, improvised responses to the moment. In the context of women's worlds, lived under sometimes harsh patriarchies in precarious times, ululation's acoustic qualities, social value, generic fluidity, and shifts in register (Gray 2016) are key to conjoining aesthetics and gendered politics.

### ULULATING TO, WITH, AND OVER SOUND STUDIES

Ululation remaps sound studies by filling out the multiplicity of sound studies narratives. Could we take ululation as a metaphor for dialogue returning amplified and inflected from the South and specified by race, gender, and struggle? That dialogue shifts the attention in sound studies to the voice as technology (Ochoa Gautier 2014) and finds sympathetic vibrations with black studies, curiously underplayed in sound studies as it is evolving.

In wanting to see more presence given to the vital genealogy of black studies, to which sound studies is indebted, I am thinking of W. E. B Du Bois ([1903] 2009) writing out from the vocables, exclamations, shouts and cries within the southern black church as the starting point for his observations on race, mobility and empowerment in the United States, or Zora Neale Hurston's primary descriptions of vocalizations in her 1930s field notes (1981), or Paul Gilroy's black Atlantic (1995) resting on sonic histories and sonic connections, acoustic traces recoverable through listening, taken up in the work of Ashon Crawley (2017), Fred Moten (2003, 2010), Alexander Weheliye (2005), and others. Could we take inspiration from black studies and from ethnographies of sound?

Could we consider ululation intervocal participation of and on behalf of the global South and, with this, on behalf of ethnography and black studies? By introducing the idea of "intervocality," Steven Feld (2012) turns from an earlier emphasis on dialogism and dialogic editing (Feld [1982] 2012) to polyphony. Dialogism approaches communication as a form of negotiation, a call and response in which the accumulating feedback shifts the terms of the conversation. By contrast, intervocality is a form of intimate covocalizing in a relational space—a space that coevolves with the vocalizing. Here

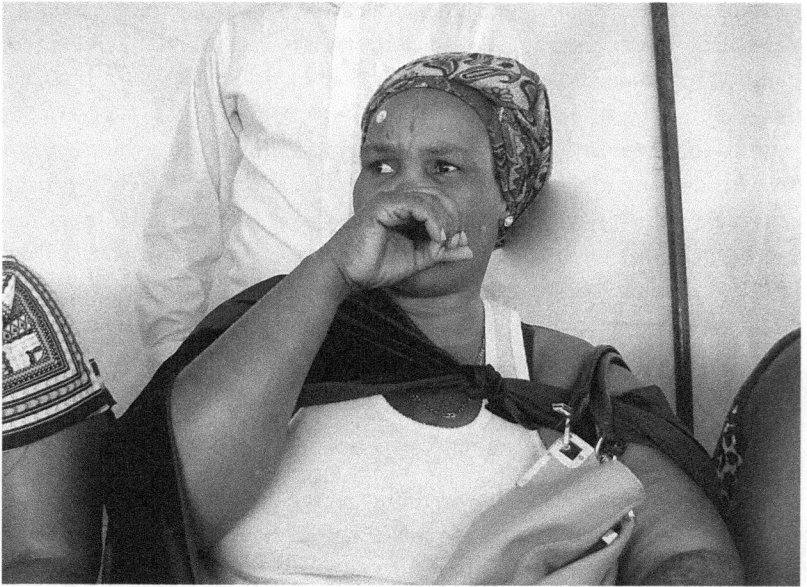

FIG 2.2. Thombilakhe Zulu ululating, December 26, 2015, Keates Drift, Msinga. Photograph by the author.

the voice is constituted through immediate improvisations that draw from a repertoire of relationships near and far, contemporaneous and historical, imagined and remembered, material and elusive, voiced and heard. Intercorporeality (Maurice Merleau-Ponty, via Fischer 2008; Weiss 1999) inflected into the voice and into listening as intervocality blends the significance of sound with its reshaping in the moment of sounding. Singular voices carry within them multiple other voices, present and elsewhere. They emerge in dialogue with others and in relation to ways of being heard (Feld 2012).

To remap sound studies, the global South, like ululation, calls for attention to its own materiality and to the particularity of its sound worlds—and thus to its histories of struggle, to an uneven interconnected terrain, to violent politics, and to the grannies crisscrossing as the wind blows.

### Notes

1. The competition is implicit: there are usually no judges, prizes, or announced winners. I elaborate the organization and aesthetic of ngoma in Meintjes (2017).

2. I transcribed dialogue and song texts from videos I recorded and translated them from isiZulu with the assistance of Siyazi Zulu. All subsequent English-language quotations were performed in isiZulu at ngoma events.

*Louise Meintjes*

3. Gena Dagel Caponi (1999) represents a variety of ways in which Gates's concept has prompted fruitful understandings of African American antiphonal musical practices. I am thinking of ululation as both double-voiced (as in signifyin[g]) and polyphonic (i.e., voicing or speaking to multiple social relationships).

4. The single article dedicated to ululation published in an ethnomusicology journal is inspired by the study of ritual language, not music, and was contributed by a linguistic anthropologist (Kuipers 1999). Improvised solicitations that are a feature of many genres of African performance share some characteristics with ululation. *Libanga* "shout-outs" in Congolese *soukous* are a crucial component of a performance, with a band member designated to improvise them (White 2008). *Mipasho* in Tanzanian *taarab* tends to be critical, and sometimes insulting (Askew 2002). In South African *kwaito*/house music, dancing fans interject vocal exclamations such as "Vuma!" ([We] agree/accept) (Steingo 2016).

5. The suffix -*ela* means "an action . . . performed for, on behalf of, or in the direction of something or someone" (Mbeje 2005: 229, quoted by Steven Black).

6. Steingo (2016) considers the relation among these three terms.

7. The historical work of Veit Erlmann (2010), Alexander Rehding (2005), and Ana María Ochoa Gautier (2014) that upend European modernity's colonial sound epistemology, and music disciplinarity is my inspiration. Likewise, the ethnography of Daniel Fisher (2016), who racializes the history of the Aboriginal Australian radio voice and proposes a "Black Pacific" that seizes on sonic blackness as a value, and Steven Feld (2012), who traces transatlantic listening of African musicians as cosmopolitans, prompt my exploration here.

8. I take this demographic description directly from Cousins and Hornby 2009.

9. This poverty line is calculated per person per month. The real 2015 value is $136 per month, converted to U.S. dollars using purchasing power parity rather than exchange rate. I thank Katharine Hall for providing these figures.

10. This figure excludes adults older than fifteen who are not economically active (e.g., students and pensioners) and is extracted from the 2011 Census by Katharine Hall (personal communication, November 28, 2016).

11. This story also appears in Meintjes (2017), where it is more extensively situated.

12. Umzansi Zulu Dancers, *Zindala Zombili*, CD, Izimpande IZI 003, Johannesburg, 2007.

13. Umzansi Zulu Dancers, *Khuzani*, LP, Gallo Music Productions MCGMP 40502, Johannesburg, 1994.

14. I am thinking here of women's vocalizations at a rain dreaming ceremony recorded by Alice Moyle in Arnhemland, Australia (Aboriginal Music, CD, Ivry-sur-Seine, France, [1977] 1992). Though not strictly ululating technique, the sound and its placement in relation to the central (men's) performance appears strikingly similar.

15. The excerpt is available on YouTube, accessed December 16, 2016, https://www .youtube.com/watch?v=VcDKLjqLUDU.

16. Ana María Ochoa Gautier contributed this point.

17. Here I follow Ellen Gray (2013), who calls attention to the entanglement of form, embodiment, and history by articulating fado "as genre," distinct from the idea of "a genre" that compels descriptive contextualization of a musical form.

18. Without materiality a voice loses its politics (Feld et al. 2004; Cavarero 2005); without artfulness the voice diminishes its narrative authority and its "capacity to claim an audience" (Malkki 1997: 223).

*References*

"All CAP/Mdukatshani Newsletters from the 1960s till the Late 1990s." 2016. *Mdukatshani*. Accessed September 6, 2016. http://www.mdukatshani.com/publications.php.

Askew, Kelly M. 2002. *Performing the Nation: Swahili Music and Cultural Politics in Tanzania*. Chicago: University of Chicago Press.

Bakhtin, Mikhail. 1986. "The Problem of Speech Genres," trans. V. W. McGee. In *Speech Genres and Other Late Essays*, ed. C. Emerson & M. Holquist, 60–102. Austin: University of Texas Press.

Black, Steven. "South African Languages." 2010. *Society for Linguistic Anthropology* blog, July 13. Accessed September 16, 2016. http://linguisticanthropology.org/blog/2010/07/13/1387.

Caponi, Gena Dagel, ed. 1999. *Signifyin(g), Sanctifyin', and Slam Dunking: A Reader in African American Expressive Culture*. Amherst: University of Massachusetts Press.

Cavarero, Adriana. 2005. *For More Than One Voice: Toward a Philosophy of Vocal Expression*, trans. Paul A. Kottman. Stanford, CA: Stanford University Press.

Cousins, Ben, and Donna Hornby. 2009. "Imithetho yomhlaba yaseMsinga: The Land Laws of Msinga and Potential Impacts of the Communal Land Rights Act." Church Agricultural Projects and Leaning and Action Project, October. Accessed July 1, 2017. http://www.mdukatshani.com/resources/MRDP%20LEAP%20research%20on%20gender%20and%20land.pdf.

Crawley, Ashon T. 2017. *Blackpentecostal Breath: The Aesthetics of Possibility*. New York: Fordham University Press.

Doyle, Peter. 2005. *Echo and Reverb: Fabricating Space in Popular Music Recording, 1900–1960*. Middletown, CT: Wesleyan University Press.

Du Bois, W. E. B. [1903] 2009. *The Souls of Black Folk*. New York: Library of America.

Erlmann, Veit. 2010. *Reason and Resonance: A History of Modern Aurality*. New York: Zone.

Feld, Steven. [1982] 2012. *Sound and Sentiment: Birds, Weeping, Poetics, and Song in Kaluli Expression*, 3rd ed. Durham, NC: Duke University Press.

Feld, Steven. 2012. *Jazz Cosmopolitanism in Accra: Five Musical Years in Ghana*. Durham, NC: Duke University Press.

Feld, Steven, Aaron A. Fox, Thomas Porcello, and David Samuels. 2004. "Vocal Anthropology: From the Music of Language to the Language of Song." In *A Companion to Linguist Anthropology*, ed. Alessandro Duranti, 321–45. Malden, MA: Blackwell.

Fischer, Sally. 2008. "Ethical Reciprocity at the Interstices of Communion and Disruption." In *Intertwinings: Interdisciplinary Encounters with Merleau-Ponty*, ed. Gail Weiss, 153–68. Albany: State University of New York Press.

Fisher, Daniel. 2016. *The Voice and Its Doubles: Media and Music in Northern Australia*. Durham, NC: Duke University Press.

Gilroy, Paul. 1995. *The Black Atlantic: Modernity and Double Consciousness*. Cambridge, MA: Harvard University Press.

Gray, L. Ellen. 2013. *Fado Resounding: Affective Politics and Urban Life*. Durham, NC: Duke University Press.

Gray, L. Ellen. 2016. "Registering Protest: Voice, Precarity, and Return in Crisis Portugal." *History and Anthropology* 27, no. 1: 60–73.

Hurston, Zora Neal. 1981. *The Sanctified Church*. Berkeley, CA: Turtle Island.

Jacobs, Jennifer E. 2007. "'Unintelligibles' in Vocal Performances at Middle Eastern Marriage Celebrations." *Text and Talk* 27, no. 4: 483–507.

Jacobs, Jennifer E. 2008. "Ululation in Levantine Society: The Cultural Reproduction of an Affective Vocalization." PhD diss., University of Pennsylvania.

Kuipers, Joel C. 1999. "Ululations from the Weyewa Highlands (Sumba): Simultaneity, Audience Response, and Models of Cooperation." *Ethnomusicology* 43, no. 3: 490–507.

Malkki, Liisa. 1997. "Speechless Emissaries: Refugees, Humanitarianism, and Dehistoricisation." In *Siting Culture: The Shifting Anthropological Object*, ed. Karen Fog Olwig and Kirsten Hastrup, 223–54. London: Routledge.

Mbeje, Audrey N. 2005. *Zulu Learners' Reference Grammar*. Madison, WI: National African Language Resource Center.

Meintjes, Louise. 2017. *Dust of the Zulu: Ngoma Aesthetics after Apartheid*. Durham, NC: Duke University Press.

Moten, Fred. 2003. *In the Break: The Aesthetics of the Black Radical Tradition*. Minneapolis: University of Minnesota Press.

Moten, Fred. 2010. *B Jenkins*. Durham, NC: Duke University Press.

Noble, Michael, Wanga Zembe, Gemma Wright, David Avenell, and Stefan Noble. 2014. *Income Poverty at Small Area Level in South Africa in 2011*. Cape Town: Southern African Social Policy Research Institute.

Ochoa Gautier, Ana María. 2014. *Aurality: Listening and Knowledge Production in Nineteenth-Century Colombia*. Durham, NC: Duke University Press.

Rehding, Alexander. 2005. "Wax Cylinder Revolutions." *Musical Quarterly* 88, no. 1: 123–50.

Smith McKoy, Sheila. 2001. *When Whites Riot: Writing Race and Violence in American and South African Cultures*. Madison: University of Wisconsin Press.

Steingo, Gavin. 2016. *Kwaito's Promise: Music and the Aesthetics of Freedom in South Africa*. Chicago: University of Chicago Press.

Steyn, Melissa, and Don Foster. 2008. "Repertoires for Talking White: Resistant Whiteness in Post-Apartheid South Africa." *Ethnic and Racial Studies* 31, no. 1: 25–51.

Vergès, Françoise. 2015. "Happiness and the Revolutionary: Singing, Dancing, and Marching." Public lecture delivered at the Johannesburg Workshop in Theory and Criticism, University of the Witwatersrand, Johannesburg, July 2.

Weheliye, Alexander G. 2005. *Phonographies: Grooves in Sonic Afro-Modernity*. Durham, NC: Duke University Press.

Weiss, Gail. 1999. *Body Images: Embodiment as Incorporeality*. New York: Routledge.

White, Bob W. 2008. *Rhumba Rules: The Politics of Dance Music in Mobutu's Zaire*. Durham, NC: Duke University Press.

*Louise Meintjes*

# 3

# How the Sea Is Sounded

## REMAPPING INDIGENOUS SOUNDINGS

## IN THE MARSHALLESE DIASPORA

*Jessica A. Schwartz*

. . . . .

It was not until long after the astronomers had begun to sound out the realms of space and to measure the distances and weigh the masses of the planets that the longing, which has always existed in the human mind to know more of the mysteries of the sea, began to be gratified. Indeed, the deep sea remained unfathomed and mysterious until after the second half of the present century had dawned upon the world; and the contemplative mariner of fifty years ago, as he looked upon the heavy bosom of the ocean and wondered at its mysteries, had nothing but myths and legends to sustain his meditation.

— G. W. LITTLEHALES, "How the Sea Is Sounded"

Thousands of years ago, when most sailors were still hugging the coast, the island peoples of the Pacific held the knowledge and skills to explore the great ocean paths around and beyond their homes. Modern instruments didn't exist — no compasses, no radio, no radar, no GPS. The Pacific peoples navigated their canoes with their own sophisticated techniques, using the seas, skies, and sea life to guide them. Their knowledge was built up through generations of experience. It was handed down through careful teaching, stories, and songs. An experienced Tongan navigator once said, "The compass can go wrong, the stars never." And that is the beauty of Pacific navigation.

— "The Canoe Is the People"

Today, approximately one-third of all Marshallese citizens live outside the Republic of the Marshall Islands (RMI), where they are denied political voice, including the right to vote in elections in their home country. They live in a transnational diaspora that often has been made marginal through imperial violence that has necessitated their out-migration. Marshallese maintain that music, memory, and self-determined movement (navigation) reproduce their cultural bonds. Marshallese in diaspora sound the sea to hear their worlds connected, to hear their "Home Sweet Home." They sound the sea to navigate the waters of our contemporary historical moment in which they are faced with extant radiological debris and other detritus of U.S. nuclear colonialism that invisibly marks their central Pacific homelands, their rich geocultural formation of dry land, water, air, and all things on and in them (the atollscape), which is usually invisible or may appear like specks of dust on the blank oceanic canvas that divides visibly "cultivated" landmasses on the globe and world maps. Marshallese sound the sea to counter this erasure and struggle against being inundated by the watery expanse—metaphorically and literally, given global warming and rising sea levels. They sound the sea as agency and redraw themselves, sounding out their networked presence across all media forms, including Facebook, with illustrations such as the one in figure 3.1, titled *Home Sweet Home*.

The illustration, from the Majolese Music Facebook page established in 2012 by Marshallese living in the United States, is an amalgam of images that evoke the place Marshallese call their "Home Sweet Home"—that is, the Marshall Islands.[1] In the top left corner is the country's flag, with its colors of blue, orange, and white. Palm trees blow in the wind, and the sun sets in a vibrant magenta glow. A bird glides in upward motion between groups of palm trees, some bearing ripening coconut fruit. The image is full of people, too, and speaks to Marshallese vitality through interconnection. The people are moving amid the waves, sailing on canoes; above the U.S. census seal, men navigate across the ocean next to a group of dolphins; elsewhere, women weave sails and handicrafts from traditional materials near modern telephone lines. At the bottom left, a woman and a man sail together. The woman looks outward, and the man smiles, humming a tune, as represented by the musical notes emanating from his mouth.

The image stresses movement, music, and community—blurring the binaries of land and sea, national and international, modern and traditional, and nature and culture on which European cartographic representations are founded and indigenous erasures are justified. The pictorial representation highlights the mobile musicality of the indigenous diaspora, a concept that

*Jessica A. Schwartz*

FIG 3.1. Art featured on the Majolese Music Facebook page in 2012.

also maintains a binary distinction in Western imaginaries of fixed or rooted indigenous people and scattered and constantly moving diasporic populations. These highly gendered representations of the diasporic male and land-based female are complicated by indigenous diasporic communications and epistemological understandings of ancestral depth as a way of moving their bodies through their terraqueous worlds. Representations of indigenous diaspora are thus political struggles—resistances—against European carto-graphic imaginaries of the natural indigene fixed on her empty land (*terra nullius*) and of the diasporic population wandering and thus not developing their land (also terra nullius), which helped justify imperial conquest and continued development projects that sustain the global North's hegemony.

Returning to the powerful symbol of the canoe and the connective waves as part of the expanse of indigenous soundings—that is, mobile music as depth—it is important to stress that, in addition to projects such as UNESCO's "The Canoe Is the People," scholars have noted the importance of acknowl-edging indigenous maritime advancements as we pursue larger projects that remap and thus redistribute resources by way of what is made sensible, au-dible, and how those sensibilities are perceived. Although we never "see" or "hear" depth measured, *Home Sweet Home* shows us how the sea is "sounded" (with "sounding" referring to the measuring of depth using an instrument) from a Marshallese indigenous epistemology, which includes logic that decenters the Western notion of sound as transcendent or as a dispersal. It returns us to the indigenous vehicle of communication, the canoe, and po-sitions it as central to a robust network of indigenous global connectivity, a remapping of the aquamarine and continental landmasses. Sound studies often treats the concept of sound from acoustical, auditory, and vibrational perspectives. The epigraphs draw on other notions of sounding and, in so doing, yield insight into different epistemologies and approaches to moving through complex worlds.

In this chapter, I unpack these seemingly unrelated concepts of sounding (in the former's sense of probing and measuring and the latter's sense of inter-generational sound-based means of sharing navigation) and articulate them to the oft-ignored cartographic and pictorial representations of the sea, or the oceanic, in sound studies.[2] By addressing sounding in these ways, I propose to remap sound studies through intersensorial understandings of deep space (bathymetrically) and deep time (indigenous relationality) to emphasize how these modes of sounding oceanic possibility are central to global movements that assert self-determination and reproduce means of sovereignty.

British maritime culture and American naval pursuits have long been associated with imperial conquest, including military actions and cultural encounters. Indigenous populations' maritime culture is less frequently explored for its political importance, rendering it part of a "disappearing" phenomenon erased by Europeans and Americans. The anxiety surrounding cultural preservation often fixes indigenous people to an imagined past,

*Jessica A. Schwartz*

forgetting about the shared developments, intercultural exchanges over time, and dynamic ways in which indigenous and nonindigenous people reproduce their worlds in intersensorial ways that often get mapped onto the visual fields by nonindigenous people and scholars. Elizabeth DeLoughrey stresses the importance of acknowledging Pacific maritime culture and the contributions made by indigenous peoples to European culture precisely through the use of their ancestral worlds, even as the imaginaries of their lands were mapped onto fixed surfaces and claimed by colonizing populations. She positions Pacific Islanders' perspectives from voyaging canoes against nineteenth-century British cartography, which, she writes, was "an attempt to inscribe and thus ideologically fix colonial territories and cultures" (DeLoughrey 2007: 194). The "flat planes" of these maps "reflects the contours of empire, and by extension, the nation-state" (DeLoughrey 2007: 194).

Colonial maps continue to define imaginaries of space and notions of mobility and access; they are mediated histories of Western epistemologies that encode cultural values of empire building. So too are the often invisible processes that ascribe or erase meaningful depth through soundings that fail to take into account networked transductions of sound symbolism and acoustical material.[3] Here, media and network studies further animate an exploration of the connective expanses and emplace indigenous peoples' contributions to and understandings of mobile media within them.[4] This chapter considers how Marshallese soundings blur lines among nation building, a politics of indigeneity, and diaspora by refusing to give way to colonial erasures of their movements—perhaps most poignantly, the silencing of the sounds that reproduce their dynamic existences. I begin with a discussion of Marshallese sounding by way of Karin Amimoto Ingersoll's notion of a rhythmic "seascape epistemology." I then offer an overview of colonial soundings that have defined and violated Marshallese people and their atollscapes (the unique geocultural imaginaries of the interconnected atoll), through which they reproduce their sounds, their cultural depths. I conclude by addressing navigational sound symbolism in the Marshallese diaspora, from the Pacific atolls to the largest Marshallese community in the continental United States (Springdale, Arkansas). I give examples of Marshallese sounding the sea to feel proximity to one other and the ancestors (home sweet home) in terms of sound reproduction, which is a form of navigating modernization and sounding out indigenous sovereignty and is symbolized by the canoe, the ocean, and sea markers such as birds. By presenting sea markers of their vital presence, Marshallese use their maritime culture to remap Western claims to cultivation and depth.[5]

The Republic of the Marshall Islands, the official name of the country that gives it international status as a sovereign state, is often referred to by Marshallese as *aelōñ kein ad* (our atolls) or *aelōñ kein* (these atolls). The etymology of *aelōñ* speaks to the interconnectedness of a seascape epistemology, which is vastly different from European derivative imaginaries of the atoll as a ring of isolated islands (Maddison 2015). One common example of a Euro-American misunderstanding of Pacific inscriptions and literacies (those practices of writing and reading ancestral land and lineage), which are verbal (oral and written) and visual, is the word for atoll, *aelōñ* (*ae* 'currents' and *lōn* 'that which is above,' such as the dry land and sky [Maddison 2015]). American missionaries, who began transcribing the Marshallese language in the 1860s, approximated the sound of *aelōñ* into the written form "ailin," a borrowing of the English word "island," which oversimplifies the Marshallese holistic concept of *aelōñ kein ad*.

The Marshallese word for sound, *ainikien*, can be read in terms of its etymological connections to other material processes of interconnection and communication, such as *aelōñ* (atoll). Like *aelōñ*, *ainikien* (sound/voice/noise) is a composite term in which *aini* is related to the words *aini/ae* (gather, gathering, collect, pool) and *ae* (current, oceanic) and *kien* refers to rules, policy, or government. When we begin from the currents and their presumed depth that humans work with rather than dominate, we can begin to acknowledge how Marshallese representations of media technologies and communications are about self-determination and sovereignty that is not overdetermined by Western epistemologies of sound as dispersal, immaterial, or isolated from other processes of movements. This logic contrasts how sound is often conceived in terms of airwaves and dispersals. We store our music libraries in the "cloud." Radio stations broadcast and reach people all over the world with their apps.[6] We are thus part of a larger broadcast culture with local variants, following early twentieth-century city planning's call to "think globally, act locally." Images of speakers and bullhorns, flaring phonograph horns, and soundwaves extending from left to right all intimate an emanating outward into the frontier of listening communities to be formed upon hearing. Such a wireless imagination in mobile music (save the wires that connect an ear-budded listener to her iPhone) plays on the notions of an individual's ability to transcend the spatiotemporal limitations of human geophysical movement. We often imagine our listening and projection as rooted in the individual broadcast, recording, or voice that is dispersed to many

*Jessica A. Schwartz*

other individual listeners who listen with their ears. Like the satellite receivers that form the tripartite processes of ground-space-ground transmission, we imagine a network of receivers and perceivers. In *The Undersea Network*, Nicole Starosielski (2015: 5) recalls a conversation with a cable engineer about the representation of digital media, about which he replied, "Satellites are just 'sexier' than cables." When thinking about cables, Western fantasies of transcendence and isolation become quickly grounded.

Marshallese navigational networks rely on Marshallese sounds that relate bodies rather than placing the navigational agency in satellites and GPS systems. These sounds become relational in the context of navigational knowledge through Marshallese soundings as *ancestral depth* in present-day maritime songs and chants. These songs and chants have specific Marshallese rhythms and shape aurality to listen for and feel the rhythms of the ocean, reproducing indigeneity as something that cannot be gendered or fixed, mapped or defined from the outside. Western scientists continue to be baffled by Marshallese navigational practices. Yet sounding the sea requires an understanding of the layered oceanic currents and current zones *and* the complex sounding practices, which include making sounds to feel proximities, that position humans rhythmically and relationality with nonhumans (animals and environmental movements).

Karin Amimoto Ingersoll develops the concept of seascape epistemology to reemplace indigenous knowledge of oceanic connectivity and fluidity that cannot be occupied, dominated, and exploited. Ingersoll's work draws on that of the Tongan scholar Epeli Hau'ofa, which imagines the sea as the place of Pacific Islanders' roots and routes (Hau'ofa 2000). Ingersoll proposes an "oceanic literacy" that is a "political and ethical act of taking back . . . history and identity through a rhythmic interaction with place: the swing of tides shuffling sand, the sharp tune of swells stacking upon each other at coastal points, the smooth sweep of clouds pulled down by the wind" (Ingersoll 2016: 129). She builds on Hau'ofa's assertion that Pacific histories cannot be understood "without knowing how to read our landscapes (and seascapes)" and maintains that this oceanic literacy is central to indigenous self-determination and "modern" identity that is "both indigenous and global" (Ingersoll 2016: 25; Hau'ofa 2000: 466).

Reading the seascape requires knowledge of the relationship between the ocean's features and island masses. Sounding the seascape often comes from those movements through it, through the atollscapes, and this can be heard in stories about navigation and the chants that prefaced and accompanied it. As the Marshallese elder Willie Mwekto (n.d.) explains, "Your language is your lamp and the light for your path." He also outlines the three languages—*lojet* (the language of the ocean), *ene* (the language of the dry land), and *lañ* (the language of the sky)—that form "the building blocks of [Marshallese] oral tradition" (quoted in Jetnil-Kijiner 2014: 60–61). While sound extends beyond humans and connects spirits of things and is made by nonhumans and nonanimals (e.g., rocks), humans have the capacity for language and the depth of language that helps care for and cross through the land. For this reason, place names are often recited to reveal the qualities and possibilities of the land. Reading practices depend on this understanding that the sounds of the three-part language, when put together orally, can reveal life potentials and connections that lie beyond immediate perception.[7]

In a compilation of Marshallese stories, Alfred Episos (2004: 31) writes, "In the old days Ri Majol used to sail; however, they chanted magic beforehand so they would know if the trip would be successful. They used magic chants to know the location of their island destination ahead of time so they would not get lost at sea once they set sail. Sometimes they became lost and could not find any atolls, but they never gave up. They kept on sailing." Chants, or *roros*, were used throughout the journey—from the building of the canoe through sailing to arrival—to lift spirits and maintain focus. Roros, like the ocean, are thought to be *m̧wilaļ* (deep) and can "reveal" specific islands and other forms, processes, and markers crucial to successful navigation. As Ingersoll (2016: 129) explains, "Rhythms don't just represent the ocean; they constitute figurative layers. Merging the body with this rhythmic sea enables a reading of the seascape's complex habits, as well as all the memories created and knowledges learned within this oceanic time and space but have been effaced by rigid colonial constructions of identity and place."

Marshallese sound the sea by sounding the rhythms of the sea, which include their ancestral bodies related to oceanic movements and sea markers named (said) through incantations that activate the body to feel, name, and move. Sounding the sea is about experiencing lived depth against Western claims to indigenous bodies as "natural" (un-historied) bodies. In this respect, Marshallese navigational chants performed through the instruments of the Marshallese human body and canoe—heard and felt—are sound reproduction. They are reproduction of journeys through space, time, and an ancestral lineage that American missionaries and colonial powers have debased and tried to silence through other sounding mechanisms.

*Jessica A. Schwartz*

The Marshall Islands consist of two parallel atoll chains (approximately thirty-three atolls and five islands, with a total of seventy square miles of land) that run from northwest to southeast across 750,000 square miles of the central Pacific Ocean. In 1787, the British sailors John Marshall and Thomas Gilbert charted what, to them, were unknown islands in the Pacific. Growth of European whaling in and fur trading across the Pacific to the Americas at the turn of the nineteenth century contributed to increased cartographic attention to the groups of islands. As Barrie MacDonald (2001: 15) explains, it was not until the 1820s, when "Adam von Krusenstern, the Russian explorer and cartographer [who] brought together all known information on the Pacific in an atlas and a series of commentaries," gave the two groups of islands their names—the Marshall Islands and the Gilbert Islands—in honor of the men who were claimed to have first sighted them.

Empire building and its uneven forms of colonial governance depended on nautical charts, maps, and instruments. It also depended on European mapping of the world as Christian and non-Christian, civilized and barbarous, owned and empty that often drew from Roman property law. Roman law and, later, European laws that enabled the growth of empire stated that land was different from the sea and the heavens, which could not be claimed by a sovereign power. Maps reflect this division of the aquamarine background and contrasting continental foreground: islands figure as barely visible dots or microscopic figures. For each colonial intervention, processes of renaming places and remapping land and politics of reproduction went together in empire building.

During the eighteenth and nineteenth centuries, the Marshall Islands received relatively limited European contact, which mostly took the form of visiting whaling vessels. That changed in 1857 when missionaries from the American Board of Commissioners for Foreign Missions (ABCFM) of Boston, Massachusetts, arrived in the southern atolls and began attempting to convert the Marshallese to Christianity. Much like their cartographic brethren who named and mapped indigenous islands, the missionaries created a Marshallese orthography by naming sounds of the Marshallese language through the sounds of English. Thinking of sounding as measurement, the ABCFM missionaries used their ears to measure the progress of moral transformation of the indigenous population as they sang Christian hymns. Through their insistence on what they considered modest dress and the practice of separating the sexes to educate them, and their disapproval of customary practices such as tattooing, chanting, and inter-atoll warfare, the missionaries—later bolstered by the Germans—altered Marshallese customary practices but did so with the acquiescence of the most powerful chiefs, who embraced Christianity and allowed it to spread.

The Germans took over political administration of the Marshall Islands following the Conference of Berlin in 1884–85, when the imperial powers divided the world into neocolonial territories and literally redrew the map of Africa and of parts of Asia and the Pacific. The Germans claimed the lands primarily to engage in the copra trade and with the approval of the Marshallese principal chief, Kabua, who received some of the profits in exchange for Marshallese labor and use of land. The Germans banned inter-atoll warfare, not out of moral inclinations like the missionaries, but because of its potential to disrupt trade. Banning this customary practice also limited Marshallese movement, which was further stymied under Japanese rule.

The Japanese annexed the Marshall Islands following World War I and another remapping of the world by the European powers based on the League of Nations' Mandate System. They blocked Marshallese mobility, particularly among Marshallese living on centrally located atolls such as Kwajalein, Jaluit, Majuro, and those farther west, such as Enewetak Atoll, when they relocated Marshallese to work alongside Koreans to build fortifications for the coming war. Prior to this, the Marshallese had retained their customary navigation practices, a highly specialized form of navigation that draws on the unique geography of the parallel chains, which enables wave piloting, a multisensorial form of sailing navigation that makes use of multiple ecological elements, such as wind, wave, and current, and animal sounds and sightings. Western scientists have been unable to fully understand Marshallese navigational practices, which the canoe pilots often depict as depending on specific sensory practices that afford a "feel" of the ocean's movements. Practitioners are authorized by the chiefs to acquire this specialized knowledge and learn through experience and intensive study of oral genres and stick charts made of palm frond midribs tied together with coconut fibers, which often include shells to represent atolls; the charts are abstract and geometrically model the motion of the currents (Ascher 1995). Canoe building and navigational precision were crucial to all aspects of life and community organization through the end of World War II, when the United States succeeded the Japanese occupation of the archipelago.

The United States—which defeated Japan in World War II and claimed various atolls, including Kwajalein, as military bases—retained possession of the Marshall Islands at the war's end and immediately imposed cultural

*Jessica A. Schwartz*

practices with often devastating consequences. The U.S. military introduced aluminum motorboats and cheap canned American food, for example, which was distributed to the Marshallese along with education in local atoll government administration, Western grade-school subjects, and American popular culture. The United States also violently inflicted ecological, biological, and cultural changes when the military conducted sixty-seven nuclear weapons tests between 1946 and 1958. Rongelap Atoll, home to one of the few remaining customary indigenous navigation schools because of the shape of its lagoon and northern location, was covered with radioactive debris from the Operation Castle Bravo shot of 1954.

From the postwar period through 1986, when the Marshall Islands achieved independence, the country was under the United Nations Trust Territory of the Pacific Islands (TTPI) administered by the United States. The TTPI was designated a strategic trust due to weapons testing, and mobility was limited. Human subjects exposed to radiation were quarantined and studied in makeshift laboratories. Bikini and Enewetak, the atolls on which the tests occurred, became "laboratories" for scientists and ecologists, who worked together to create new notions of systemic connectivity and isolation based on the atoll isolate (DeLoughrey 2013). Marshallese people were not allowed to leave the immediate region except to go to high school (often on the island state of Chuuk) or college, into the military service, or, accompanied by doctors appointed by the Atomic Energy Commission, to laboratories in the United States for further research on the effects of radiation on humans. There are many songs that describe painful isolation from ancestral homelands and work through tensions that emerged when customary ways of life were forcibly rerouted through medical procedure, such as thyroid surgery conducted on Marshallese citizens who had been exposed to radiation but did not fully understand what had caused their disease. Marshallese were given unfamiliar terminology to make sense of their increasingly foreign encounters with their own lands and bodies and thus drew from the familiar to re-emplace themselves and mitigate disorientation. The metaphor of the canoe and the journey through the currents, as distance markers and gravitational forces, are often used in these songs.[8] The songs often consist of Marshallese lyrics set to borrowed (perhaps modified) hymns, whaling songs, and German, Japanese, or American tunes.

In 1986, the United States signed into law the Compact of Free Association (COFA), the political treaty with the RMI. The compact is an experiment in international law, a product of U.S. decolonial efforts that retains structural inequalities and, thus, neocolonial opportunities. In legal terms, an "association" is a formal link between two countries that have unequal power. The associate (here, the RMI) is recognized as having "significant subordination and delegation of competence" to the principal nation (here, the United States). The COFA, in the most basic terms, enables Marshallese citizens to travel, work, and live in the United States without a visa or permit. The COFA also established, among its many sections, Section 177, which explains the terms of nuclear compensation and espousal, as well as the establishment of the Nuclear Claims Tribunal.

As mentioned earlier, approximately one-third of the Marshallese population live outside the RMI, where, because they are Marshallese citizens, they have no political voice and cannot vote in either RMI or U.S. elections. The disenfranchisement of Marshallese citizens by disallowing absentee voting was approved by legislative vote in the Nitijela in October 2016. Following the vote, the Council of Iroij (customary chiefs) made an appeal to Nitijela Speaker Kenneth Kedi requesting that the law be reversed so Marshallese citizens could vote wherever they reside. At the time of this writing, this is still being debated. Regardless, the move has angered many diasporic Marshallese and plays into tense debates on the past, present, and future of the islands that are resisting mass migration, even as global warming and sea level rise continue to make the ancestral lands uninhabitable. The restriction discriminates against some Marshallese who send resources back to the islands in the form of money, goods, and even a radio station. The decision participates in power distributions that further define who has a voice and what that voice sounds like in contemporary Marshallese society, and this is contrary to Marshallese trends in fashioning what they consider new islands or atolls in Middle America.

One problem, then, is the different definition of "island" as connected by routes or "island" as isolate (remote utopia, prison, or petri dish). Another problem is based on contested definitions of freedom as a joint effort or based on the individual's being able to exercise her free will within a liberal economy. American colonial interventions that pushed ideologies of free will and created the "petri-dish" condition of atomic laboratories in the Marshall Islands are constitutive of voters' rights, which is in contest with the sea-of-islands mentality. Yet if voices are votes, which are routed in and through the land, they also seem to be a protective ("sound") measure against further expropriation from the American mainland in a historical moment when many Marshallese have serious concerns about anthropogenic climate change. Marshallese voices, the canoe, sea markers, and watery bodies themselves are symbols of indigenous self-determination and cultural strength.

*Jessica A. Schwartz*

They are the means of indigenous navigation through interconnections of sea waves and the airwaves, bodies across time and space. The efforts to revitalize navigation after World War II and the symbolic use of voice, canoe, sea markers (birds), and watery bodies remap the world according to Marshallese sounding practices (gathering and circulating, *ainikien*).

NATIONAL TELECOMMUNICATIONS

The revitalization of navigation and canoe building taking place within Marshallese national formation and post–Cold War politics of indigeneity is addressed not only to tourism or appeals to legitimacy, although these are important components. It is also crucial for Marshallese in remembering their forms of communality and intergenerational connections that are based on deeply composed oral genres.[9] Against Western-centric notions of monetary inheritance and individualism, Marshallese can then engage in the epistemic practice of *jitdaṃ kapeel*, or learning about their land and lineage by asking the elders. Since the entire community works together on customary projects, such as canoe building, launching, and navigation, preserving the culture means having the opportunity to talk, sing, chant, and share values that are often in contrast with the hegemonic culture, especially in external diaspora. Marshallese customs, or what they term *mantin Majōl*, afford opportunities for dynamic, relational space of intervention against fragmentary social dispersal. They enable Marshallese to approach modern technologies with their values and expectations for communal engagement, which include teaching and learning through legends, as well as the sounds of the Marshallese language.

As the nation worked to define itself politically, a push was made toward cultural revitalization. This revitalization took place in tandem with postwar developments in infrastructure, including media and communications. As Marshallese increasingly moved to Hawai'i and the continental United States, accumulative sound forms—including those associated with new technologies—were playing an important role in cohering communities. Interestingly, the Marshallese perspective on sound and gathering (*ainikien*) has worked to gather not only people but also these network foundations, giving them Marshallese heritage. One example of this is the National Telecommunications Authority (NTA) cell cards that, during my fieldwork in 2008–2010, one had to purchase to make cell-phone calls or use WiFi, which, before fiber optic cable was laid in 2010, was comparatively quite slow.

FIG 3.2. Cell phone card from the Marshall Islands National Telecommunications Authority.

The angular sail shapes at the left of the image are repeated, which is part of Marshallese visual and sound patterning. Sounds are repeated within a word to signify ongoing activity. The water, either the lagoon or the ocean, is present in both, reminding the learned viewer of the original communication lines among islet, islands, and atolls — namely, the waves and currents of the sea and the air through which markers, connections, are revealed. People are also present. The images here are of contemporary *riwut* (model outrigger canoes) that are raced.

The NTA seal is located in the upper left corner. The top of the exterior circle reads "Marshall Islands," and the bottom reads "National Telecommunications Authority." Inside the circle, there is a national twenty-four-point star, which represents the twenty-one atoll municipalities, with four longer rays symbolizing the nations "sub-centers" (Majuro, Kwajalein, Jaluit, and Wotje). A ground satellite points toward the star. Under it, a landline telephone hovers around the rippling water. The telephone and the satellite form the shape of an outrigger canoe with sail. Unlike representations of satellites aiming at outer space (which leave the orbiting satellite and receiver ground satellite out of view), this satellite is in communication with the guiding star, much as a navigator on a canoe would have been. The wavy water line connects the two islets, both of which have palm trees towering over all other objects below and reaching the star above. However, the island on the left-hand

*Jessica A. Schwartz*

side, where the tree is farther away from the star, has a towering building, which represents either the capital or a location outside the Marshall Islands' outer islands. Much like the legends, this visual inscription archives customary mobile communications alongside wired/cabled and electronic networks and the content afforded by them.

The monetary value of the card is shown on the right-hand side. Money is seen as one of the most insidious American imports that stands in contrast to *mantin Majōl* values based on (1) respect for the land and iroij (matrilineal ties); (2) subsistence (*ejjelok wonen*, "free of cost, everything is from the land"); and (3) togetherness (*ippān doon*). In contrast to the numerical value, the pictures represent those values just by showcasing the sails, the canoes, the water opening outward, and the people working on the canoes together. Respect for the chief is also written into these activities, and this respect is an extension of the matrilineal inheritance.

### BIRDS ON THE GROUND IN "SPRINGDALE ATOLL"

*A New Island: The Marshallese in Arkansas* is a film directed by Dale Carpenter with guidance from and narrative by RMI Consul-General Carmen Chong-Gum, a Marshallese woman.[10] Carpenter, a white male documentary filmmaker and professor of journalism at the University of Arkansas, proposed the title, which was ear-catching because it juxtaposed the notion of an "island" with Arkansas, a land-locked state situated in the southern region of Middle America. Chong-Gum liked the title and worked with Carpenter to create a storyline that brought cultural awareness to non-Marshallese. The film begins with a roro (chant) performed by the Marshallese elder Simon Milne. It is juxtaposed with images of Middle American urbanity and the poultry industry in which many Marshallese residents work. The documentary offers viewers an intergenerational journey of out-migration, and it decenters masculinist notions of imperial governance and domination of the feminized seascape by men. *A New Island* is itself a sonic remapping of Northwest Arkansas, and this sounding, which layers the depth of Marshallese culture onto a terrain known more for "woo pig sooey" than for being culturally diverse, should remind us of the connections and geopolitical divisions of these settlers' negotiating their "new islands." A part of the story of the Marshallese journey to the midwestern United States centers on the bird and the poultry industry.

Navigational roro often reference birds, which were used as sea markers. Different types of birds feed at different spots in the ocean, and by revealing which type of bird one sees, the land mass will be known. Also, birds were an important part of Marshallese diets to supplement protein from fish and other seafood. Today, Marshallese eat chicken, which is mostly imported from the United States. Marshallese also work in poultry plants, such as the large Tyson plants in Northwest Arkansas. The Marshallese word for bird is *bao*, and chicken is *bao in lal* (bird on the ground). Songs often imitate the *bao* sounds, and it is preferred when a woman's voice soars like a bird in flight. There are many legends that have to do with bird-related transformations. This speaks to the metaphor of the bird in flight for various types of motion. The bird on the ground, the chicken, today represents economic subsistence, which is necessary but also feels stifling to many Marshallese. Moreover, the American way of life is dependent on money, so movement is dependent on money. People are trained to sit and work for eight hours a day, five days a week, with minimal breaks, only to have to pay to exercise, especially if it is dangerous to exercise in public. This is anathema to Marshallese journeying, a concept called *jambo*.

Media often share the stories of the Marshallese in Arkansas—for example, an Associated Press story that features Daisy Loeak, a Tyson poultry plant worker responsible for assessing the "freshly-killed Cornish hen that comes down the production line to decide if it's of premium quality" (Associated Press 2015). The story tells of her labor: "She routes the hens onto conveyor belts before they're packed into boxes and flash-frozen. Out of 300 workers at the plant, Loeak is one of about 120 Marshallese. She moved to Springdale in 2008 with her grandparents, who traveled to the United States for a funeral and ended up staying." Loeak is part of the latest generation of workers who come to landlocked states for employment, education, and health care at a cost of living lower than that on the Pacific Coast or in Hawaiʻi. Other Marshallese enclaves, in Oklahoma, Missouri, and Kansas, are growing due to the availability of these types of poultry and other production line jobs. Approximately eight thousand Marshallese live in Springdale, and as many as fifteen thousand reside in the Midwest. The story continues by pointing out the Americanization of Marshallese names and the Marshallese naming of American places. "It's Chickendale, not Springdale," said Loeak, whose birth name is Daisina but who adopted "Daisy" because it is easier for Americans to pronounce. She wells up with tears as she talks about rising sea levels and says she misses her homeland. "In the Marshall Islands, it's just more carefree," she said. "You go where you want" (Associated Press 2015).

However, Marshallese who have been forcibly relocated because of nuclear testing, such as the Bikinians, do not feel that the Marshall Islands

*Jessica A. Schwartz*

allows them to go anywhere. They feel trapped on the islands to which they have been relocated: Kili Island, a lone island without a lagoon that is less than a square mile in area, and Ejit Island in Majuro Atoll, which is about the size of a football field. Many of the displaced Bikinians live on Kili Island, which they refer to as "the rock" and "jail." Bikinians pride themselves on being skilled navigators, and Alson Kelen, the former mayor of Bikini who now lives in Majuro Atoll, the capital, started the nonprofit Waan Aelōñ in Majōl (Canoes of the Marshall Islands; WAM) with the American canoe researcher and developer Dennis Alesso; WAM teaches canoe building and navigation and offers a contemporary approach to Marshallese vocational schooling as an alternative to college.[11] Marshallese education creates dialogic space in which a new model of jitdam kapeel is practiced, as per the motto of the College of the Marshall Islands—"Jitdam Kapeel"—and its logo, an outrigger on which a Marshallese family waves under the RMI flag sail. Waan Aelōñ in Majōl encourages Marshallese forms of education to increase physical, intellectual, and socioeconomic and mobility. The program is involved in youth outreach to mitigate the damage done by alcohol, drugs, widespread unemployment, poverty, and familial estrangement, as evidenced by the unprecedented homelessness in the capital city, which, even at one or two people, is for the Marshallese unacceptable.

Marshallese existence is founded on having a home, a land tract and ties that form social organization and coherence. This is why not being able to safely inhabit their ancestral lands has devastated Bikinians. The image of the grounded chick, separated from its mother and unable to feed itself (jojoḷāār), often appears in contemporary Bikinian creative expressive practices. More than a thousand Bikinians have relocated to the Midwest, and in 2016 a branch office of the Bikini-Kili-Ejit local government opened in Springdale. Unlike in Kili and Ejit, where the Bikinian population is centered, in Springdale and surrounding communities the homes of Bikinians (like those of other Marshallese atoll populations in diaspora) are spread out as social organization in the United States continues to change. While familial ties to atolls remain strong, church membership now is one of the primary means of sociality and group membership.

As discussed earlier, the German colonial administration and the ABCFM actively discouraged Marshallese customary practices and expression, which included many sound-based practices, such as roro and aje drumming, and reduced others. The missionaries' spelling of aelōñ, the Marshallese word for atoll, as "ailin" obfuscates the Marshallese origins of the term and its relational meaning based on wave piloting. There are many words that exemplify the reductive nature of the missionaries' orthography. Marshallese continue to debate whether the new orthography that addresses the sounds of Marshallese and, using the International Phonetic Alphabet, aims to approximate them phonetically or whether the missionary orthography is better for customary practices. While proponents of the newer orthography (used in the *Marshallese-English Dictionary* [Abo 1976]) argue that the older orthography was developed by Americans with untrained ears, Marshallese who value the old spelling feel that the new version allows Americans to imagine that they have easy access to the language without an understanding of the land. In addition, the notion exists for some that the spoken language activates other customary knowledge and bodily experience of the words and that the written language is the language of the Bible and hymns, and should remain so.

Regardless of their preference, Marshallese used the forms and structures given by the missionaries, such as the church building and musical form, to retain their skills, language, and different grammars of cooperation. When the United States brought motorboats, it also brought money, food, and a new notion of leisure time. Church festivities grew in size, and church song competitions grew more elaborate, archiving customary motions and chants in what is known as the *jebta* (songfest competition, also sometimes spelled *jepta*). The jebta events are rigorously rehearsed in the RMI and in diasporic communities during the Kūrijmōj season, which runs through December, and are performed on Christmas. In Springdale, Consul General Chong-Gum decided to introduce a jebta competition into an event that would further Marshallese-Arkansas cultural diplomacy and understanding.

The songfest competition necessitates that Marshallese learn songs, dances, chants, and stories from other Marshallese — even from some in the RMI who travel to Arkansas to teach them. Jebta themes vary, but common performances enact canoe building and fishing or imitate birds in flight. Some of the movements have morphed into lassoing motions and shouts of "yee-haw" from the audience that resound the regional culture.

According to Chong-Gum, the jebta dances are actually a series of actions that are "translated" from "what [the dancers] are saying. First it has to be the lyrics. Then when they are making movements, they are giving actions to the words." She discussed how the movements can vary between groups, even if they are based on the same theme, such as fishing:

> Depending on the culture . . . , there are different ways of fishing:
> dive, throwing a net, or gathering, . . . or especially on the ocean side,
> when the water comes in it's high, they create a swamp, [and] when

*Jessica A. Schwartz*

FIG 3.3. Jebta performers compete during the "Battle of the Jepta" event in Springdale, Arkansas. MEI Photo/April L. Brown.

the tide goes out, the fish are stuck. [She mimes picking up a bunch of fish in her arms.] They gather them up. So there are different ways of fishing, so depending upon what kind of fishing they are using, they are making the movements according to it. It's about context. . . . [The dancers] are [also] trying to create movement that no one has come up with, so creativity is a part of composing. Not just composing, but we have to go down to the root of it, what they are singing, and the dance will reflect the lyrics. So . . . years from now, maybe [what the dancers show is] not going to be running after the pigs or after the chickens, . . . and trying to kill them in certain ways, but they'll be performing dances about how they work at Tyson Foods, about getting the chicken ready . . . [and] having Tyson in the lyrics.[12]

This explanation of jebta cuts through the arguments about preservation and change that often plague sound scholars. Chong-Gum's statement reminds us that fidelity to sound traditions is located less in the sounds themselves, as reproducible, than within the aggregate community work that goes into and is generated from the sounds. Decontextualized sound, such as the chants that are written in the TTPI's (and my) notebooks, falls flat without the material world to activate the meaning that the language reveals. Moreover, much of the grammar of such language-based revelations is meant to

*How the Sea Is Sounded*

accommodate the spirits it activates. For example, while I was sharing a song with an elderly Marshallese couple, there was a noticeable flattening of the melody. To my ears, this part echoed a roro, but they told me that it was an invitation for another person to chant. Certain strands of ethnomusicology and sound studies have at their roots (and routes) this central fear of the disappearance of indigenous cultures, as Roshanak Kheshti (2015) explores.[13] While the phonograph was meant to mitigate imperial damage that both disparaged indigenous cultures and diminished them through cordoning, sickness, and cultural conversion practices, at times, as Kheshti notes, it has created the conditions that reproduce imperial fantasies via the recorders' aural inscriptions.

Such recordings also create an archival imagination that becomes part of what Elizabeth Povinelli (2002, 2011) has explored as "the crisis of legitimacy," in which indigenous populations must make themselves legible within the confines of dominant culture. Here, it is not the process of sounds that have accumulated ancestral voices and knowledge as a practice of jitdam kapeel and are thus awaiting a response but, rather, the sounds themselves that are judged as "music" and recorded for the purposes of maintaining "tradition" (i.e., loss and the creation of archives as colonial accumulation) or for "entertainment" (i.e., disposable and capitalist accumulation). Jebta offers dynamic preservation. In Marshallese modes of sound reproduction, the dances are repeated exactly if the situation warrants. Rather than maintaining an anxious attachment to origins from which claims to authenticity (and resources) are often made, the jebta accumulates by bringing together education, entertainment, and culture — spheres that have become separate in America and, by extension, in Marshallese culture (and acoustemological orientations). This jebta event in Springdale is held to collect money to advance educational pursuits among the Marshallese so they can have skills to participate in a networked world. This is the notion of *anemkwōj* (freedom) in association with the United States that is infrequently realized because of a mix of racism, negative press, lack of English skills, and the maintenance of tightly knit Marshallese communities that place familial obligations above work and school. Jebta events show how even the chicken can be contextualized to provide different types of mobility that ultimately may obviate the need to include Tyson in future jebta.

Jebta songfests are compelling to watch because of the rigorous dancing (or *piit*, from the English "beat"), the inclusion of Marshallese chants and songs, and the militaristic leader who directs the lines of male and female teenagers while blowing commands on the whistle in sharp eighth-note

*Jessica A. Schwartz*

bursts. The song form is "sticky" in this respect, and because it aggregates cultural flows materially rendered intelligible as song and dance in bodied motions and voices, it connects people. In fact, the competitions are judged by equal numbers of white American educators and Marshallese elders, which brings its own set of power dynamics, misunderstandings, and difficulties. The point is that Marshallese imagine their songfests as mobile music with challenges that are not transcended just through performance. The American teachers and the Marshallese elders are selected as judges to promote intercultural connections, and they judge based on intersecting values that they discuss, as well. The songfest is an opportunity for communication between these two parties that might not otherwise exist. The event thus becomes a node, and the conversation becomes a connection that extends to the next node, and so on.

### SOUNDS OF THE ISLANDS

At 6 AM Central Standard Time on August 21, 2015, the Marshallese radio station KMRW 98.9 FM in Springdale, Arkansas, began its first broadcast. In the previous year, KMRW—founded, funded, and managed by the Marshallese music enthusiast Larry Muller—had grown from a prerecorded offering of Marshallese-language music streaming on smartphone and tablet apps to a 24/7 broadcast with regular live programming on the schedule. With an influx of Marshall Islanders to the region, and with V7AB, the main radio station in Majuro, the capital of the RMI, broadcasting (on the radio and streaming via Android app) only from 6 AM to 11:30 PM (Marshall Islands Time) and encountering inoperative periods due to rusted equipment, KMRW has grown out of necessity. Marshallese began calling Muller from the Marshall Islands with special requests; that is how he learned that V7AB's transmitter had rusted and fallen into the water and that the station was broadcasting over an FM frequency from the airport, meaning that parts of Majuro Atoll and the outer islands could not receive the station. Muller says that people call in to make song requests and announcements from the RMI, and that people call in from all over with birthday wishes and greetings. "Now [people] call to say 'Happy Mother's Day' from Arkansas to the Marshall Islands," Muller said, emphasizing that the messages are going from Arkansas to the Marshall Islands and not the other way around, indicating a change in tides and flows.[14]

To some extent, KMRW's programming follows the Marshallese intergenerational communication model. Musical programs are geared toward elders and adults who remember the atolls; when these programs activate their memories, the elders can teach youth about the islands and their customs. The station also broadcasts weather forecasts; ads for local businesses, such as George's Chicken; the "Voices of the Marshallese Church" program on Sundays; and "Kejro im Manit Eo Arro" (You and Me and Our Culture), an hour-long broadcast on Friday evenings in which Marshallese elder Franklein Henos discusses proper Marshallese pronunciation (e.g., the differences between the Ralik and the Ratak chain dialects) and customs (e.g., how to approach a chief). Because this is the sole educational show aimed at youth and disperses knowledge widely, which in itself falls outside Marshallese custom, it is not universally supported. Some Marshallese disagree with the accuracy of Henos's content, for instance, while others dislike that he is broadcasting cultural information at all over a public medium.

Muller was trained as a sailor and employed as a captain when he lived in the Marshall Islands. When one sees navigation as a Marshallese sound reproduction technology, however, the transition from captain to radio station manager seems less abrupt—especially since waves and training to recognize wave patterns are more widely available in the islands than in Middle America. The composition of new islands requires currents to signal proximity, which radio stations can offer—whether via broadcast or via an app. The radio station's location and content provide the route, because, as in navigation, the route is dependent on the human's feeling the oceanic activity through the movement of the canoe. Revealing markers through the incantation of deeply composed, contemporary transportation vehicles and media are forms of material movements that connect people physically, a point that is made abundantly clear in an essay by the radio station manager Antari Elbōn (2004: 145), who writes,

> The Marshall Islands has many means of communication. The field-trip ships sail on a monthly basis. The Ministry of Resources and Development sends Marshall Islands Marine Resource Agency (MIMRA) boats to Arno twice a week and elsewhere irregularly. The Rongelap community now has a live-aboard dive boat. . . . Planes fly. . . . The privately owned cable television on Majuro has 21 channels. People on Majuro and Kwajalein use electrical power, and people on outer island[s] use diesel generators to show videos. The newspaper . . . [arrives] once a week. Only V7AB . . . reaches our people 8 hours a day, every day of the week.

FIG 3.4. Henson Abon and Michael Capelle of KMRW 98.9 FM, in an
advertisement on the station's Facebook page.

Elbōn begins with vehicles and ends with the radio. While the image of
the radio broadcast signal or a conical speaker often connotes the possibility
for sound to travel through dispersion, Marshallese continue to represent
musical mobilities as transport through combined ocean and airwaves and
currents rather than simply representing the radio as a transcendence of
time, space, or place. The image in an advertisement for KMRW, 98.9 FM —
"Sounds of the Islands," the only non–Christian Marshallese music station in
the United States — recalls the previous representations of Marshallese communications possibilities, with its land, trees, and body of water stretching
out beyond the borders of the picture. Marshallese often refer to the Ozarks
as "the mountains," and, while this picture is not of the Ozarks (it is most
likely the Rockies), it speaks to the imaginary of Marshallese material possibilities, especially since the atolls are being threatened, and damaged, by
anthropogenic climate change.

The advertisement for the station, which began broadcasting in 2015,
is the embodiment of Marshallese and American value systems and their
acoustemological orientations. On the left, Henson Abon, a veteran broad-
caster from V7AB in Majuro, stands without pretense wearing a nautical-
theme button-down shirt. Behind him is an image of a palm tree that serves

as background to the microphone that disperses sound beyond the bounds of the FM frequency strength and broadcast area. The images float on water in the ad, and, as in the other images (figures 3.1 and 3.2), the men are a crucial component. The younger man (right) represents U.S. currency by wearing a money-patterned sweatshirt, and with his hand in a Hawaiian "hang loose" gesture, he recalls the larger Pacific diaspora of which his generation was part in terms of decolonization movements and pan-Pacific (and, ultimately, pan-Pacific and Caribbean) modes of cultural and regional belonging that increasingly became reflected in Marshallese music. The picture of material abundance that is represented in the ad is re-sounded in the continuous stream, the flow, of music back to the RMI, forming imaginaries about life in diaspora that shape arguments about the continued relationship among the rights to land, voice, movement, and community.

## THE POLITICS OF SOUND REPRODUCTION

The erasure of the islanders and their cultural knowledge has not been without issue, which includes claims to customary power and notions of what is "properly" Marshallese. As we have seen, the cutting of communication lines—the songs, the throats, the boats, and so on—has increased power for some. Knowledge about navigation, canoe building, and traditional medicine continues to be guarded and only selectively passed down, which is disconcerting both for those who want the knowledge shared or archived because they are concerned that it will be forgotten and for those who insist that the best way to maintain traditional culture is to retain its customary practices of exclusivity.

For the Marshallese, the pursuit of self-determination means different things to different people, depending on where they stand in the customary hierarchy, their educational background, and their socioeconomic status. Sound reproduction as political reproduction depends on methods of media network archaeology that consider the roots, and the shape of the routes, through which sounds as revealing journeys are aggregated into acoustic assemblages. Sound reproduction in diaspora is therefore a working composition bursting with unresolved dissonance, literally, in the unheard voices on the islands that represent uncounted votes in elections; in the challenges to a radio personality's show; and in the seemingly fluid transition between the bird in flight to the grounded bird when judges' discussions unravel in cross-cultural miscommunication. The politics of sound reproduction under the COFA are the politics of cultural reproduction or, to put it another way, the

*Jessica A. Schwartz*

biopolitical controls within demanding geopolitical networks that aggregate power by way of sonorous dispositions and intelligibilities. A tension exists between the rigors of sound production and what is "proper" in Marshallese culture, or a collection of sounds that are materially meaningful because they reveal other collections, or accumulations, of materials that have been passed down through specific customs.

The arguments I have shared in this chapter, of course, ultimately concern political power. Sound reproduction involves cultural reproduction and the reproducibility of political clout. In other words, discussions of what constitutes "proper" sound collection and media communication in Marshallese reveal specific modes of power accrual, from chiefly wars in canoes to colonial treaties and imaginaries of islands as isolated places. Even nuclear weapons testing and anthropogenic climate change have been profitable for some (as the Associated Press report obliquely suggests), and for many islanders, the compound fear of being detached from their land and from their lineage permanently, without any benefits or sense of belonging, is always present. Moreover, colonial culture created the conditions where the voices and sounds of group-based inheritance became articulated as (individualized) property. Under the logic of neoliberal capitalism, anxieties surrounding dispossession of customary land exist and are played out through sounded journeys that reinforce the lineage-based hierarchical power of accrual and thus the possibility of divestment of collectivized agency. What is being sounded, of course, are the tensions of modernization and the larger economic forces that, conveyed through liberal language, promise freedom as individualized upward mobility and dematerialization—notions which contradict customary Marshallese notions of freedom in terms of interdependent mobility and matrilineal kinship. These freedoms, in their tensions, resound connections between dispersed listeners who continue to listen to Marshallese radio apps and hear elders tell stories that reproduce sounded journeys in canoes through the water and birds in flight. The radio waves and the oceanic waves that circulate voices and bodies speak to the contested freedoms of a Marshallese customary hierarchy that resounds through the voices in the Council of Iroij and the collectivized movements of bodies in the Marshallese diaspora.

*Notes*

Epigraphs: G. W. Littlehales, "How the Sea Is Sounded," *Popular Science Monthly* 44 (January 1894): 334–41; "The Canoe Is the People," Local and Indigenous Knowledge Systems, United Nations Educational, Scientific, and Cultural Organization, 2017, accessed May 14, 2017, http://www.canoeisthepeople.org/index.php. The quote from the Tongan navigator is from Lewis 1999: 56.

1. The free social-media platform Facebook, with its homepage, video, and chat functions, has become a popular means of communication and expression, connecting Marshallese within the United States and those abroad with Marshallese in the home islands.

2. There is a foundational contrast in Western and indigenous imagination of what constitutes depth, how depth is measured, and how depth is useful in these movements. A sounding is the use of an instrument to measure the depth of the sea, but it can also be a probing, like that of a medical instrument or an interrogation. In the nineteenth century, echo sounding became commonplace when telegraphy made advances in sounding machines to aid in mapping the ocean floor for navigation. Piano wire and, later, sonar were used to transmit sound (vibrational) pulses into the water. While they are thus articulated to one another through practical utility, the etymologies of the English words "sound" show that they come from different roots: (1) *sonus* (the auditory sense of hearing noise or tones); and (2) *sund* (to measure the depth of a body of water; swimming or movement through the water), which also means "secure," "whole." Sounding is measuring and a means of quantifying distance via an instrument that renders the distance through vibrations, such as the vibration of a string.

3. Jonathan Sterne's statement that "modern technologies of sound reproduction use devices called transducers, which turn sound into something else and that something else back into sound" (Sterne 2003: 22) has left a generous space for the interpretation of what "modern technologies of sound reproduction" might be. Imperial mapping could be part of a modern technology of sound reproduction, given the inscriptions of sounds that are gathered and circulated in various soundings of people, places, and histories. These inscriptions of sounds of names onto maps from nautical data are, in Ana María Ochoa Gautier's term, "acoustic assemblages," which collate different translations of hearings and listenings to be reheard, relistened to, and transduced from inscription to sound all over again (Ochoa Gautier 2014). Ochoa Gautier was interested in how indigenous and Afro-descended people were heard and transcribed by European colonials and how these hearings (mishearings) became part of a process of colonial control and, eventually, nation building (through the transformation of voices). Her work helps us reconsider how maps, mapping, and acoustic assemblages participate in remapping sound worlds and, thus, sound studies.

4. I refer to "media archaeology" as "a way to investigate the new media cultures through insights from past media" (Parikka 2012: 2). "Network archaeology," as coined by Starosielski (2015: 15), is an approach that works to "historicize the movements and connections enabled by distributions systems and to reveal the environments that shape their contemporary media circulation."

*Jessica A. Schwartz*

5. Epeli Hauʻofa's (2000) seminal work, "Our Sea of Islands," connected the "roots" and "routes" of Oceanic peoples and questioned imperialist perspectives of Pacific islands as small and isolated. Steven Feld (1996) has written about "flow" and water-falls in Kaluli acoustemologies.

6. Although conceptualizations of radio stations differ across cultures, as Noriko Manbe explores in an essay on online radio, I include this idea of the "radio broadcast" here to conjure ideas of the broadcast as a dispersal of music and sounds—a sending out of streamed music in which only some types of intervention by the listener are possible (see Manabe 2014).

7. For more information on place names, see Feld and Basso 1996; Weiner 1991.

8. I explore these songs in greater detail in Schwartz, *Radiation Sounds: Marshallese Music and Nuclear Silences* (forthcoming).

9. "Deeply composed" is a term I have been working with to think through the concept of ṃwilaḷ (deep) in Marshallese music. It is based on place names and what has gone into naming a place. Through its ancestral work with the land, some of the language in some of the songs is "deep." Drawing on the concept of through-composition, then, I term such practices "deep composition" or "deeply composed."

10. Dale Carpenter, dir., *A New Island: Marshallese in Arkansas*, film, University of Arkansas, Fayetteville, 2004.

11. Each atoll is a political municipality, with senators and a mayor.

12. Carmen Chong-Gum, e-mail correspondence, August 22, 2014.

13. Certain strands of ethnomusicology and sound studies draw from acoustic ecology, or soundscape studies, which was pioneered by the composer and environ-mentalist R. Murray Schafer and others at Simon Fraser University in Vancouver, British Columbia. Soundscape studies were part of the World Soundscape Project, which aimed, in an era of emergent northern hemispheric environmentalism, to find a certain sonorous global harmony and preserve sounds of a world threatened by loss (see Schafer 1993). For a robust critique of this model, see Ochoa Gautier 2016.

14. Larry Muller, e-mail correspondence, May 13, 2017.

## References

Abo, Takaji, Byron W. Bender, Alfred Capelle, and Tony DeBrum. 1976. *Marshallese-English Dictionary*. Honolulu: University of Hawaiʻi Press.

Ascher, Marcia. 1995. "Models and Maps from the Marshall Islands: A Case in Ethno-mathematics." *Historia Mathematica* 22: 347–70.

Associated Press. 2015. "Arkansas a Refuge from Rising Seas in Marshall Islands." Fox News Science, November 27. Accessed October 30, 2106. http://www.foxnews.com/science/2015/11/27/arkansas-refuge-from-rising-seas-in-marshall-islands.html.

DeLoughrey, Elizabeth M. 2007. *Routes and Roots: Navigating the Caribbean and Pacific Island Literatures*. Honolulu: University of Hawaiʻi Press.

DeLoughrey, Elizabeth M. 2013. "The Myth of Isolates: Ecosystem Ecologies in the Nuclear Pacific." *Cultural Geographies* 20, no. 2: 167–84.

Elbōn, Antari. 2004. "Heartbeat of the Marshalls." In *Life in the Republic of the Marshall Islands, Mour ilo Republic*, ed. Anono L. Loeak, Veronica C. Kiluwe, and Linda Crowl, 143–48. Suva, Fiji: Institute of Pacific Studies, University of South Pacific.

Episos, Alfred. 2004. "Bwebwenato, Legends." In *Life in the Republic of the Marshall Islands, Mour ilo Republic*, ed. Anono L. Loeak, Veronica C. Kiluwe, and Linda Crowl, 31–40. Suva, Fiji: Institute of Pacific Studies, University of the South Pacific.

Feld, Steven. 1996. "Waterfalls of Song: An Acoustemology of Place Resounding in Bosavi, Papua New Guinea." In *Senses of Place*, ed. Steven Feld and Keith H. Basso, 91–136. Santa Fe, NM: School of American Research Press.

Feld, Steven, and Keith H. Basso, eds. *Senses of Place*. Santa Fe, NM: School of American Research Press.

Hau'ofa, Epeli. 2000. "Epilogue." In *Remembrance of Pacific Pasts: An Invitation to Remake History*, ed. Robert Borofsky, 453–72. Honolulu: University of Hawai'i Press.

Ingersoll, Karin. 2016. *Waves of Knowing: A Seascape Epistemology*. Durham, NC: Duke University Press.

Jetnil-Kijiner, Kathy. 2014. "IEP JĀLTOK: A History of Marshallese Literature." MA thesis, University of Hawai'i, Manoa.

Kheshti, Roshanak. 2015. *Modernity's Ear: Listening to Race and Gender in World Music*. New York: New York University Press.

Lewis, David, and Hekunukumai Busby. 1999. "Voyages." In *Proceedings of the Waka Moana Symposium 1996: Voyages from the Past to the Future*, ed. Hans-Dieter Bader and Peter McCurdy, 13–31. Auckland: New Zealand National Maritime Museum.

MacDonald, Barrie. 2001. *Cinderellas of the Empire: Towards a History of Kiribati and Tuvalu*. Suva, Fiji: Institute of Pacific Studies, University of the South Pacific.

Maddison, Benetick Kabua. 2015. "Oral Histories of the Marshallese Diaspora: Translation as Cultural Conversation." Paper presented at the American Historical Association, Pacific Coast Branch Conference, Sacramento, CA, August.

Manabe, Noriko. 2014. "A Tale of Two Countries: Online Radio in the United States and Japan." In *The Oxford Handbook of Mobile Music Studies*, 2 vols., ed. Sumanth Gopinath and Jason Stanyek, 1:456–95. New York: Oxford University Press.

Mwekto, Willie. n.d. "Life in Those Days: Interview by Newton Lajuan," trans. Andrea and Terry Hazzard. *Marshall Islands Stories*. Accessed October 20, 2016. http://mistories.org/life-Mwekto-text.php.

Ochoa Gautier, Ana María. 2014. *Aurality: Listening and Knowledge in Nineteenth-Century Colombia*. Durham, NC: Duke University Press.

Ochoa Gautier, Ana María. 2016. "Acoustic Multinaturalism, the Value of Nature, and the Nature of Music in Ecomusicology." *boundary 2* 43, no. 1: 107–41.

Parikka, Jussi. 2012. *What Is Media Archaeology?* Cambridge, UK: Polity Press.

Povinelli, Elizabeth A. 2002. *The Cunning of Recognition: Indigenous Alterities and the Making of Australian Multiculturalism*. Durham, NC: Duke University Press.

Jessica A. Schwartz

Povinelli, Elizabeth A. 2011. *Economies of Abandonment: Social Belonging and Endurance in Late Liberalism*. Durham, NC: Duke University Press.

Schafer, R. Murray. 1993. *The Soundscape: Our Sonic Environment and the Tuning of the World*. New York: Inner Traditions.

Schwartz, Jessica A. Forthcoming. *Radiation Sounds: Marshallese Music and Nuclear Silences*. Durham, NC: Duke University Press.

Starosielski, Nicole. 2015. *The Undersea Network*. Durham, NC: Duke University Press.

Sterne, Jonathan. 2003. *The Audible Past: Cultural Origins of Sound Reproduction*. Durham, NC: Duke University Press.

Weiner, James F. 1991. *The Empty Place: Poetry, Space, and Being among the Foi of Papua New Guinea*. Bloomington: Indiana University Press.

# PART II

MULTIPLE LIMINOLOGIES

.....

# Antenatal Aurality

# in Pacific Afro-Colombian Midwifery

*Jairo Moreno*

.....

*To Ana María Ochoa Gautier, for opening ears.*

## INTRODUCTION

The practices of Afro-descendent midwives in the Afro-Pacific region of Colombia (hereafter, Pacific Afro-Colombia) gather a complex material and spiritual network linking body, health, life, and death.[1] They hold immeasurable social value; to be born in the care of a midwife means to be born in the community. Midwives also have considerable powers, possessing a gift (*don*, in Spanish) and being capable of inscribing specific characteristics in a newborn that will shape her or his place in the community—for example, the length at which a midwife cuts the umbilical cord determines the degree of masculinity or femininity of a newborn (Arocha Rodríguez 1999; Losonczy 1989). Above all, for the community the midwife's first auscultation of a fetus's heartbeat affirms a life defined as an aurally relational mode of existence. Life, in this relational sense, is what happens between her ear and the child's heartbeat.

But life is also a biological matter. Listening plays a central place in midwives' science. In their subsequent clinical diagnosis and monitoring practice, midwives learnedly perceive biological indexes of life and fetal well-being, with particular attention paid to the acoustic subtleties of heartbeats (Moreno and Steingo 2018). Their audile practices affirm and sustain antenatal vitality in and for the community.[2]

And there is another, spiritual, dimension to life. Antenatal aurality partakes in fundamental ways in the wagers their community makes on a specific divine configuration of the human. In this sense, as Hugo Portela Guarín and María Elvira Molano (2016: 10) write, midwives constitute "the sacred link between human life and nature" and mediate the domains of the living human and the divinities that the not-yet-born directly embody. There is good reason for this, as another audile perspective holds place of privilege in antenatality: early on during pregnancy, fetuses have the capacity not only to hear across the liquid and permeable milieu in which they dwell—which is also well-known in biomedical research—but also to listen. This is a terrifying power in the case of children who die during childbirth or shortly thereafter, given their ontological proximity to divinities. Their listening is active for seven days after death and is feared because they can report back to divinities what they heard during their brief coexistence with the living (Losonczy 1989).

That midwives in Pacific Afro-Colombia are practiced phenomenologists, and their attention to sound indicates their attunement to the biological dimension of human life, to that which is held to be essential and specific to human constitution (gestating bodies as organisms of a particular kind, with certain anatomical features and faculties, and so on): they listen to and provide a link to "nature," as Portela Guarín and Molano remark. That they do so in a milieu in which "beliefs" about human origin link fetal aurality with divine powers suggests their dynamic participation in a particular auditory culture, performing the world-making capacities of human collectives everywhere: they listen culturally. In sum, we might say that midwives listen to the universality of nature from the particularity of Pacific Afro-Colombian culture. And they do this from a practiced intersectionality in which dictates from biomedical institutions, where they are required to receive basic aid training, and esoteric knowledge (from the point of view of healthcare officials) takes place without apparent contradiction.

This neat bifurcation between nature and culture and between biological fact and cosmological, spiritual, or metaphysical belief, however, has as a precondition a modern ontology grounded by what Philippe Descola (2013) calls "naturalism." Under this ontological regime, humans differ from themselves along a temporal continuum and on the basis of gradual stages of development by which, during the passage from ante- to postnatality, humans attain greater "reflective consciousness, subjectivity, and ability to signify, and mastery over symbols and the language by means of which [they] express those faculties" (Descola 2013: 173). In turn, "human groups distinguish

*Jairo Moreno*

themselves from one another by the particular manner in which they make use of those aptitudes by virtue of a kind of internal disposition that we now prefer to call 'culture'" (Descola 2013: 173).

How then to mediate those aspects of Pacific Afro-Colombian midwifery that share naturalism's (and biomedicine's) understanding of human bodies as occurring within a single developmental vector of gestation and the presumed linear ontogenic unity of the human—the faculty of hearing included—on the one hand, and, on the other, claims about antenatal aurality that appear to reconfigure this vector, reverse the notion of audile development, and assign capacious faculties to fetuses and newborns as an expression of a wholly "practical metaphysics"?[3] In one possible answer, midwives carry the responsibility to socialize cosmology and to "cosmologize" nature, so to speak, and they do so at the symbolically powerful threshold of human antenatality, birth, and postnatality. But there is more. I approach this apparent duality as something other than an absolute separation between physics (nature) and metaphysics (culture)—and something other than a culturalist "deconstruction" of modern nature, or, even worse, an impossible reconciliation of diverse views on a unitary nature from a multicultural position. Instead, I propose to understand antenatal existence as a set of ontological variations on the "human" given in and through the aural perspective linking (and separating) the human and the para-human.[4] I give this phenomenon the name "quasi-life," a life that is fully in its own and at a given moment but that coexists both with its negation—that is, death—and with the possibility of transcending both life and death. As I argue, in Pacific Afro-Colombian midwifery naturalism plays a relatively small part; it shares and is contested by other ontological arrangements, particularly "perspectivism" and "analogism," each of which expresses a particular set of modes of identification and relation, of mapping interiorities and exteriorities, temporalities, spatializations, mediation, categorization, and figuration (Descola 2013: 115). The ontology of life, then, emerges from the assemblage of these arrangements, with sound and listening playing a particularly important role, as I show.

The argument for this ontological set of variations in antenatal aurality requires some methodological wagers, which I make by bringing together ethnography and ethnology alongside philosophical speculation, each drawing from different configurations and conceptual frameworks. The political stakes of this wager and their potential for cartographies of sound studies are sketched in the final section of the chapter. First, I arrive at an account of midwife auscultation through ethnographic work with Mrs. Sixta

Tulia Zambrano Cuero y Caicedo, in dialogue with (so-called global North or Western) phenomenological approaches to sound studies. A more general analysis, under the notion of quasi-life, emerges in conversation with ethnological work on Amazonian and Amerindian conceptions of death and anthropological work at the regional level of the Pacific Afro-Colombia, and with proposals by continental philosophers who, with an implicit naturalist understanding of life, have nonetheless engaged the ambiguities inherent to human life and death. The wager on comparativism here is partly heuristic, indeed, expressing a desire to bring things into dialogue and contradiction, but it also represents an effort to experiment with a notion of remapping that cannot be only the inclusion of unheard of sounds and modes of listening or the inversion of so-called Northern presumptions about sound. But I am getting ahead of myself. First, some background on the ground on which Mrs. Zambrano works.

.........

Mrs. Zambrano, born in 1949, first aided in childbirth when she was thirteen.[5] Originally hailing from Calabazal, a small seaside settlement in the northern part of Nariño Province, in the municipality of El Charco, on the Pacific coast of Colombia, she has since 1974 lived and worked in the port city of Buenaventura (roughly two hundred kilometers north of Calabazal as the crow flies but more than a day away by boat, the only means of travel along the coast and its many fluvial settlements inland). Buenaventura (population 415,000) is Colombia's largest port and the site of perhaps the worst urban violence in the country, condensing armed conflict between right-wing paramilitaries associated with drug trafficking, remnants of leftist guerrillas, militarization by the state, and extreme socioeconomic disparity emerging from the privatization of the port.[6]

By Mrs. Zambrano's account, she has assisted more than one hundred births. The tools of her trade are standard among midwives in the zone: first, the Pinard horn, a trumpet-shaped, wooden auscultation device with which she listens to the heartbeat of the fetus; second, local medicinal herbs (*nacedera, carpintero morado, chilca* and *ruda, hojita de la virgen*); third, a cured bottle (*botella curada*) in which medicinal herbs are mixed with *viche*, a potent homemade alcoholic distillation made from sugar cane and said to cure intestinal worms and snake bites as much as to promote fertility and enhance sexual vigor. With a smile, she remarks, viche comes in handy to relax expectant mothers but also to render them less chatty, the better for her to hear the subtle, often faint sounds of the fetal heartbeat. As a set, these tools denote the

*Jairo Moreno*

key role that listening and sound play in antenatal care. In turn, the centrality of listening and sound forms part of sensorial approaches distributed between hearing and touch, mainly, but inseparable from the agency of plants and vernacular distilled beverages, in this case, in the constitution of the network of technics (i.e., technologies and techniques) assembled under the practice of midwifery.

Mrs. Zambrano has been a member of the Asociación de Parteras Unidas del Pacífico (Association of United Midwives of the Pacific; ASOPARUPA) since 1998, when the group was founded by Rosmilda Quiñones. Today, more than 250 midwives work with the association.[7] In October 2016, the Colombian Ministry of Culture's Division of Patrimony included Pacific Afro-Colombian midwifery and its associated knowledges (*saberes asociados*) in the Lista Representativa del Patrimonio Cultural Inmaterial del Ámbito Nacional (Representative List of Immaterial Cultural Patrimony within the National Sphere [Colombian Ministry of Culture 2016]).

The association's self-proclaimed mission is "to preserve ancestral cultural knowledges and the practice of traditional medicine together with scientific knowledges, fomenting unity, solidarity, and interculturality to remain always in search of knowledges that might help us complement and articulate the sense of belonging and communitarian commitment that animates us to live in the Colombian Pacific, since we are agents for the support of life."[8] Part of a larger epistemic moment in the recent history of Colombian governance, ASOPARUPA belongs to the burgeoning social movement networks in the region that in part seek to harness blackness (*Afro-descendencia*, in Spanish) to state recognition (Escobar 2008; Restrepo 2013). It also belongs to a general moment in which Pacific Afro-Colombians have become increasingly self-conscious about deep temporalities associated with their place in the wider African diaspora, signaled here by the imperative "to preserve ancestral cultural knowledges." Speaking to this recent recognition, the Fundación Activos Culturales del Afro (Foundation for Cultural Goods of the Afro; ACUA) stated that "midwifery is perhaps the most ancient tradition in humankind" (*la partería es quizá la tradición más antigua de la humanidad* [Colombian Ministry of Culture 2016]). Liceth Quiñones, the daughter of Rosmilda Quiñones and current head of ASOPARUPA, pointed to the cultural and territorial identities and the particular focus on the politics of gender, race, and class in Colombia as singular dimensions of midwifery (Colombian Ministry of Culture 2016). For the Ministry of Culture, the next step is setting up a dialogue with the proper channels that mediate biomedical practices (the Ministry of Health) to have the medical establishment recognize midwifery. At present, ASOPARUPA midwives must attend basic health training at the main local hospital, but Mrs. Zambrano observes that doctors there, while sometimes respectful of her practice, do not regard midwifery as on par with biomedical obstetrics and gynecology. In the midst of these force fields and in the face of unrelenting urban violence and precarity in the city of Buenaventura, Mrs. Zambrano and her colleagues carry on with their work.

## TOWARD A VERNACULAR PHENOMENOLOGY OF ANTENATAL AUDITION

Mrs. Zambrano listens intently, trying to locate the heartbeat of the fetus.[9] The mother, meanwhile, averts her gaze, at first unable to listen in to a life that she sustains but that in a radical paradox audibly experiences her every sound. Although the Pinard horn renders Mrs. Zambrano's hearing monaural, it helps gather the complex sounds occurring in the womb and those issuing from elsewhere in the mother's body. Placental blood flows, vascular murmurs, and gastrointestinal gurgling sounds and growls fuse together in cacophony alongside respiratory whirls and the unavoidable thump of the mother's heartbeat, usually assertive and precise in its inexorable percussion. Amid this corporeal multiphony Mrs. Zambrano gently moves the horn over the mother's abdomen, trying to get as close as possible to the fetus's actual heart, the better to hear what she describes as "a soft, muffled drumming sound accompanied by a gentle breath-like hush," the sound of a heartbeat.

Achieving this level of auscultatory proficiency takes considerable effort and experience. Mrs. Zambrano's auscultation practice is not fundamentally different from that of Western biomedicine: bodily sounds index physiological information, helping to monitor well-being and, when irregularities are detected, to diagnose possible problems (Rice 2013; Sterne 2001). Similar to biomedicine, the listening practices and audile techniques of midwives follow a classic phenomenological scheme bringing together an intentional listening subject, an object of listening, and a means to bracket off whatever is not conducive to the purest identification of the sonic object. Listening entails deixis and denotes a particular location and aural perspective.

Most basically, clinical auscultation is akin to reduced listening, aurally searching for the best possible definition of a single sound within an otherwise complex sound field. Like a "tonic sound" (Chion 2016: 56), the fetal heartbeat, as faint as it may be, is brought into relief thanks to the functioning of a human ear that seeks *form* (e.g., pulsation, periodicity, continuity, and regularity) and *materializing sound indices* (e.g., the muffledness of

*Jairo Moreno*

the fetal heart that Mrs. Zambrano describes) (Chion 2016: 57, 103). But the vagaries of auditory perception are such that, as Michel Chion explains, it is impossible to *completely* exclude a "sound from the auditory field"; there is no frame for sound, no borders delimiting, structuring, or enclosing sound (Chion 2016: 27).

Mrs. Zambrano, of course, proceeds pragmatically, engaged in the kind of intentional listening concerned with the location and provenance of the heartbeat. As she puts it, "One listens around the belly, looking for the child's heart until it is audible but being careful not to push too hard into the mother's abdomen and provoke reactions from the child, who does not want to be bothered." Materializing indices "lead [the] sound back to its cause," writes Chion (2016: 103), and combined with the form of sound, they gather as the criteria and listening schemata that allow her to determine the well-being of the fetus. In strict terms, sound can never be mono-causal, but for the purposes of monitoring and sustaining life, for Mrs. Zambrano, the sounds that she alone hears issuing from the mother's womb, sending covibrational impulses up through the Pinard horn, directing waves toward the tympanic membrane that are then transduced into the electrical currents of her auditory system, *are* the sound *of* the heart. And this heart, for humans and in accordance to biomedicine, *is* the index of life's existence. But it is something else besides.

### THE EVENT OF THE SONOROUS

Thus far we have a rough sketch of the combination of reduced and causal listening of fetal heart monitoring. In even stricter terms, we would add that Mrs. Zambrano attains a "sonic *source* object" (Chion 2016: 105, emphasis added)—the heart—but not the heartbeat as sound *object*. This refusal of sound as an object reflects a strong correlationalist position according to which "sound is not graspable outside of a dialectic between the place of the source and the place of listening" (105). The notion that sound can be an object always entails its reduction to morphology or materiality. And even there, the substance of this object would be actually "its modulations . . . palpitations . . . kinetic curve . . . [and] information" and so on (204). All these remain temporally and spatially variable within the same "sound." So short of refusing even the analytic potential of the reification of sound as object altogether, we arrive at the conclusion, with Chion, that sound is name of the *contradiction* between its object-like and non-object-like characters (210).

As an alternative to the notion of the sound object and as an expression of its irreducible temporality there exist proposals to conceive of sounds as events in which, as the analytic philosopher Casey O'Callaghan (2009: 28) puts it, a "moving object disturbs a surrounding medium and sets it moving." O'Callaghan works with "the intuitive notion of events as particulars which take time and may or may not essentially involve change" (36n9). From another philosophical corner, Jean-Luc Nancy invokes an eventality of the sonorous, placing, however, the accent on "listening." Listening, he affirms, is "first of all presence in the sense of a present that is not a being (. . . intransitive, stable, consistent . . .), but rather a coming and a passing, an extending and a penetrating" (13). For Nancy, "sonorous time" is not a "simple succession"; it is "a present in waves on a swell, not a point on a line . . . a time that opens up . . . that envelops or separates . . . that stretches out or contracts" (13). For his part, Chion (2016: 40) remarks on how certain things "such as respiration (ebb and flow of the sea) have the ability to render duration unreal, making it escape linear time." This expresses the fact that "sound . . . unfolds and manifests itself *in* time, and is a vital process, an active energy" (Chion 2016: 187).

Despite their differences, these philosophical accounts of the event of sound and listening share a mindfulness of temporality. First, sound and listening constitute events to the degree that they "take time" (O'Callaghan) or unfold "in time" (Chion) or in which "time opens up" (Nancy 2007). Second, certain things (respiration, ebb and flow of the sea, and almost certainly the rhythms of a heartbeat) present a singular configuration such that they disrupt a presumed linearity of time, sound constituting, for Nancy, a particular time: "sonorous time."

This "present in waves of a swell" of sonorous time carries ontological consequences for Nancy. It gives "birth to presence" as unbound and always relational (Nancy 1993) and corresponds to his understanding of resonance as a back-and-forth referral and deferral that discloses being itself as a resonant sharing.

While in fundamental agreement with this ontological outlook, the insistence on "event" has everything to do with its corrective appeal against the more static "object," of becoming over being, as so much contemporary thought insists. But in the practical metaphysics of Pacific Afro-Colombian midwifery, some things cannot be framed in terms of an either-or of the event. Instead, they need to be understood within a both-and structure tied to basic forms of existence that go by the name of life and death. We may temper the philosophers' metaphysical enthusiasm by convoking Elizabeth

*Jairo Moreno*

Povinelli's notion of the quasi-event ("a form of occurring that never punc-
tures the horizon of the here and now and there and then and yet forms the
basis of forms of existence to stay in place or alter their place") to "focus our
attention on forces of condensation, manifestation, and endurance rather
than on the borders of objects" (Povinelli 2016: 21). That is, we may move
elsewhere than the phenomenological ontology of the event of sound and
listening, but without disavowing its insights, to consider how actual "forms
of existence stay in place or alter their place." The term "endurance" captures
not only the ongoing struggle of midwives for a place within the health structures that organize Colombian biopolitics, but also the very real fact of mortality and risk. In addition to the particular eventality of sound and listening, there are forms of cultural survival that are inseparable from particular
perspectives on "nature." Let us, then, mind the particular perspective from
which, in the Pacific Afro-Colombian region, the not-yet-born and the recently dead newborns listen.

### QUASI-EVENT AND QUASI-LIFE

The anthropologist Eduardo Restrepo explains how, in Pacific Afro-
Colombian communities, the "worlds of the living and the dead are insepa-
rably and indelibly connected. The dead, saints, and visions dwell among the
living" (Restrepo 2002: 14; see also Restrepo 2011). For example, when a person dies, her soul or shade (*sombra*) leaves the body. During the wake, staged
at home for nine days, a pot with water is left for the shade to quench its
thirst. Children occupy a distinct though not unrelated space. When a child
dies before age seven, she is regarded as having died without a soul, although
she is held to have had a spirit. This condition makes of a dead child a "little
angel," guaranteeing that she will rejoin the divinities in heaven, not having
had the need to atone for a fallen and sinful soul. In heaven, she will intercede
with the divinities on behalf of her parents and godparents. In the complex
funerary rites for a child, called *chigualo*, joy is the predominant emotion.[10]
Other, material elements figure forth—for example, tears avoided in order
not to drown the child. As Restrepo remarks, this funerary rite bridges the
divine sphere with the world of the living, throughout which circulate divine
entities and those free of sin, such as the dead child.[11]

Death does not constitute merely a biologically defined state, an end point,
or the ceasing otherwise of life. Rather, it is the becoming nonhuman of the
human through the transformation of spirit into a quasi-divinity. It is a kind
of reinstauration of the status held during pregnancy—a status, as I have

suggested, that is considered ontologically higher than living humans yet ontologically lower than divinities. This becoming Other of the Same is taken as an opportunity to obtain concessions from divinities: the newly dead intercede on behalf of the living. At the same time, death, from the *corporeal* perspective of the living, is a radically final event. Thus, as part of a chigualo, the mother of a deceased infant is left to grieve alone during the final hours before the child's body is interred.

The question here is this: what is death (and life) such that there are lingering resonances among beings across either side of their divide? The concern with birth and with its apparent obverse, death, constitutes an unarguable "universal" across human cultures preoccupied with defining borders that, despite the corporeal (and, let us say, "natural") finality of death, never appear clearly enough to the living and yet offer the surest index of human finitude. These borders, such as they are, demand complex metaphysical maneuvers from thinkers concerned with the interrelation between death and life.

Jacques Derrida articulates this in a productive way. "In death one waits for this other (myself) that never arrives (there) together," he writes (Derrida 1993: 65). For the living, death is a kind of impossible becoming possible: "an impossibility that can nevertheless *appear* or announce itself *as such*, an impossibility *whose appearing as such would be possible* . . . an impossibility that one can await or expect, an impossibility the limits of which one can expect or at whose limits one can wait" (Derrida 1993, 73). That this impossibility appears *as such* contaminates life in the figure of the one who is coming (*l'arrivant*) — that is, "one can await or expect death." This is not merely a matter of anxiously awaiting death. Because of the temporal spacing that it introduces to life, this anticipation becomes a matter of survival. As Martin Hägglund (2008: 1) puts it, life is defined "as essentially mortal and as inherently divided by time[;] to survive is never to be absolutely present." In short, Derrida places death as a constitutive spacing at the very heart of life's temporality. One is not surprised, then, with Derrida's name for this life that is, besides "life," the awaiting for death:[12] "life/death" (*la-vie-la-morte*).

"Life/death": this figuration demarcates a distinction between life and death at the same time that it blurs their boundaries by juxtaposing temporalities. My interest here is in how life/death in its very occurrence — in its eventality — happens both partially and as a totality that, one might say, exceeds it. The event of life/death, in short, is a *quasi-event*; it is an event that is not quite one but, rather, multidimensional and irreducible, that in its very occurrence compels a blurring of the boundaries of its happening.[13]

*Jairo Moreno*

The quasi-event life/death divides the living human from itself even as it seeks to release traditional notions of life from its boundaries, bringing life and death into a kind of resonance, not dissimilarly from the sound event discussed earlier. But three things put a damper on my conceptual enthusiasm. First, the Derridean quasi-event thinks death from the seemingly obvious perspective of life: only the living can think their death. Second, despite its dislocation and redoubling of the living self, this quasi-event takes as its analytical basis the individual. Third, for all intents and purposes, life/death operates with the background of a biological—that is, naturalist—notion of death against which the philosopher makes an ethical and a political claim: first, death entails radical responsibility toward this Other that always will have arrived, and second, no life is ever sovereign. And so, although elegant and conceptually pliable, this philosophical formulation operates as a metaphysical response to the uncontestable fact of human biological death.[14] These two domains—metaphysics and "biology"—remain separate and have no possibility to produce something other as a result of their intertwining, to say nothing of the impossibility of ever regarding sensory perception of any kind to be a concern: the living individual may await death, but this will never engage "listening." The ethical and political potential of life/death is limited precisely by the very separation that the notion seeks to disrupt and by the individual character it presumes. The individual constitutes an effect of the naturalist commitment to the notion of unique interiority for humans and between humans.[15]

The example of funerary rituals and dead infants in the Pacific Afro-Colombian fold together metaphysical and biological dimensions in a practical metaphysics that are ethical in how they attend to the constitution of the Other (divinities and quasi-divinities) that people there constitutively are—or once were, as part of the principle of species individuation they uphold. These metaphysics, too, have a political dimension in how these ethics activate principles of inclusion and exclusion in which some existents belong fully or partially in the community or exceed the community and in how the dead themselves decisively participate in these distributions of what is or is not shared. Let us further consider why this is so there, for these people, and why antenatality and postbirth mortality might compel the kind of ethics and politics they do.

Death haunts antenatality. "Anything can happen to the child," as Mrs. Zambrano remarks. But there is more. Antenatality shares with the quasi-event of death a complex temporal (and ontological) arrangement, as well as a distinctively aural character. In her ethnography of black-indigenous relations in the neighboring province of Choco, north of Buenaventura, the anthropologist Anne-Marie Losonczy (1989: 49–50) observes that "conception is a 'divine matter' that comes from Above and therefore introduces God as mediator between man and woman [and] positions the origin of children in another place: extra-human but wholly positivized." Partly transcendent and partly immanent, antenatal temporality extends across divinities, the not-yet-born and, in the figure of the mother and the midwife, the living. The not-yet-born directly incarnates divine power, which renders her metaphysical and physical. At the same time, conception, pregnancy, and birth all participate in the process of species individuation that makes humans the kind of being that they are. As Losonczy (1989: 50) continues, "This origin [of children] is positioned in the same space as that of the divine creation of the first men." With this in mind, my first proposal is to double up on the ontology of death and to think of antenatality as a quasi-event and of the life that takes place therein as a *quasi-life*.

This quasi-life, however, presents challenges to the community. Indeed, as Martha Cecilia Navarro Valencia (2012: 179–80) reports, in Pacific Afro-Colombia "giving birth has been related historically with the unpredictable and mysterious." This mysteriousness has as its background a particular aurality.[16] Not quite a divinity, but higher up than already born humans, the not-yet-born child possesses the gift of hearing. By this, they do not mean only that, as is now well known by biomedicine, the inner ear is fully "developed" by the twentieth week of pregnancy, hearing being the first sense faculty to form, or that after twenty-eight weeks there is consistency in "blink-startle responses to vibroacoustic stimulation, indicating maturation of the auditory pathways of the central nervous system," as described by the American Academy of Pediatrics (1997: 724). Rather, this capacity to hear is also the capacity to *listen in* to the living. Aware of this protean aurality and of the extra-human character of the fetus, the living exercise caution around pregnant women. Hearing persists after death—what I call quasi-death. In the event of death, the child continues to listen, and greater precautions are taken, since the creature has already communed with the living and taken in their language. People consider the acquisition of language coterminous with gaining "experience" in life, which has as its correlate the relative contamination of a being formerly pure, at birth and before. Here communication is a main concern: the divinities attend to the child directly but listen to the living only indirectly, presupposing that the fetus, the newly born, and the recently deceased infant are not just immersed in social relations as

*Jairo Moreno*

much as fully developed people are but also actively and decisively partake in a network of aurality. Social relations—speech and listening—rather than naturalist intrahuman distinctions, constitute the relations among humans, recently deceased infants, and divinities. Each of these, however, marks a distinct ontological location, what I have alluded to as a set of ontological variations aurally engaged.

Communication forms part of a space of aural interdictions and precautions. In the Pacific Afro-Colombia, analogies link virtually everything, which is seen, for instance, in the compatibility among humors, substances (e.g., medicinal plants used by midwives), and binary classifications of the cold-hot type that govern human physiology. In analogism, the potential interchangeability of everything provokes "fear of being invaded by an intrusive and alienating identity," which, in turn, entails "a need to keep workable and efficient channels of communication open between all the parts of all beings and to maintain the many circumstances and influences that ensures their stability and proper functioning" (Descola 2013: 226). The listening dead child, more than an intrusion, constitutes a potential extrusion whose rendering of earthly foibles to the divinities can bring instability to the community. Similarly, precautions about the acoustic environment of the antenatal child suggest how this entity remains vulnerable, no matter its semidivine status, to disruptive intrusions.

### SPECIATION, DIFFERENCE, AND CONTINUITY

Antenatality constitutes a set of beginnings, we might say, sonically given in the capacity of hearing and listening that refers to the myth of human origin in Pacific Afro-Colombia, according to which divinity is the common fund of humanity. Adapting Christian tenets, humans appear both in God's image and different from Him. This involves (1) ontologies of speciation; (2) principles of differentiation among humans and between humans and divinities; and (3) relations of continuity and discontinuity within and among the spiritual ("interior") and corporeal ("exterior") dimensions of humans, not-yet-born humans, and divinities.

We may take heed of Eduardo Viveiros de Castro's diagram for Amazonian peoples' speciation as a point of departure: "the narrativization of the indigenous plane of immanence articulates in a privileged way the causes and consequences of speciation—[that is,] the assumption of a specific corporeality—by the . . . actants therein, all of whom are conceived as sharing a general unstable condition in which the aspects of humans and nonhumans are inextricably enmeshed" (Viveiros de Castro 2014: 65). In other words, the undifferentiated and pure multiplicity that constitutes the "place of immanence" gives way to a first moment in which, before being humans and nonhuman animals, all creatures shared in their humanity and possessed a soul, as well as the capacity to communicate among themselves.[17] In this precosmological condition, the "corporeal and spiritual dimensions of beings do not yet conceal each other" (Viveiros de Castro 2014: 66). According to myth, differences among entities appeared when some failed to heed the warning that specific calls by other entities needed to go unattended (Lévi-Strauss 1983: 147–63)—an aural relation, in short, caused the disruption of the first moment. At this point, in Viveiros de Castro gloss, animals became "animals," retaining, however, the interior qualities of soul (their humanity) but gaining different physical corporeal characteristics. As a result, animals see themselves as human but see humans as animals, while humans see themselves as human and animals as animals.

Because both types of creatures retain the dual character of their prior condition, each can experience its Other through what Peter Skafish calls "translational means," such as Shamanism (Viveiros de Castro 2014: 12). This perspectivist arrangement entails an intensive (e.g., intraspecies) and an extensive (e.g., interspecies) differentiation in which all beings remain metaphysically contiguous in virtue of possessing a generic soul but may be physically discontinuous by virtue of the specific bodies they wear, a "specific corporeality," as Viveiros de Castro puts it. In the ontological arrangement that Descola terms "animism," animals and humans share their interiority by virtue of the soul that all have, and they differ with regard to their exterior.[18]

What, then, of Pacific Afro-Colombia, where we are considering ontological variations of a single species—the human? And where, in Descola's account, does a scheme of shared interiorities and exteriorities link everything together in an endless web of analogies (which he calls analogism)?

Let us take the case of antenatality. Since we are dealing with human life, antenatality constitutes the appearance of living humans from a quasi-life that is a corporeal preformation of the already born human and at the same time a spiritually superior entity in relation to the already born human. In their mythological condition, the not-yet-born are both finite and infinite, they self-differ and embody "an intensive superposition of heterogeneous states rather than an extensive transposition of homogeneous states" (Viveiros de Castro 2014: 66). The temporal implications are significant and belong in the world of what we might call paradox: "Each mythic subject, being a pure virtuality, 'was already previously' what it 'would be next' and this is why it

*Jairo Moreno*

is not something actually determined" (2014: 67). The yet-to-be-born child embodies this paradox.

My proposal is that the antenatal and the postmortem designate something like *intensive differences* in relation to postnatal humans, all entities ultimately being expressions of divinity. Juxtaposed to these intensive differences, the aural perspectives that each entity has on the other operate in the manner of *extensive differences* that each gains from the specific corporeal position she or he occupies and possesses. What I have called a set of ontological variations encompassing antenatality, postnatality, and postmortality (of infants) reveals something of the prior condition of shared divinity as a matter of aurality and communicability. In other words, given the radical interrelation among each part of the set (antenatal, postmortem, postnatal, divinities), the ontological task of antenatal aurality, so to speak, becomes one of marking off discontinuities among parts that otherwise lie along an ontological continuum.[19]

That hearing and listening should be the mark of the quasi-event of antenatality and the guiding index for marking the differences and similarities among humans, not-yet-born humans, and divinities invites the question of possible continuities and discontinuities within and between spiritual and corporeal dimensions. Pacific Afro-Colombian antenatal aurality refuses psychic (psyche in the sense of soul) and metaphysical discontinuity between the human and the divine, or, more specifically, between quasi-life—and, indeed, quasi-death—and the divine. The corporeal-incorporeal distinction between quasi-life and divinities is less decisive for their constitution here than the respective positions they occupy in relation to one another as listening entities and, in the case of quasi-life, as an entity capable of communicating. Similarly, as embodiments of humanity vis-à-vis divinity, a human life (i.e., an already born and grown person) and a quasi-life emerge and are discernible as the particular forms that they are out of the aural position each one has on the other.

Aurality, then, is a mark of a differentiating relationality, not as a representation in sounding form or as an idea of listening or as a mediation between the biological and the spiritual. In addition to helping reveal prior intensive differences, this relationality is fraught with danger and risk as much as it is with connecting and communicating. Its ethical dimension remains in place—each part attends to the others and cares for them—but politically the aural configuration at play here signals a radical lack of sovereignty of postnatal humans and the predominance of a strict hierarchical ontology in which they (we) occupy the lowest rung.

It would not be incorrect to place the general field of aurality discussed within an acoustemology of antenatal existence, "the agency of knowing the world through sound," in Steven Feld's pithy definition (Feld 2012a: 49). After all, Mrs. Zambrano's work, for example, is quintessentially a practice of acoustic and sonorous understanding and knowing, a mode of grasping meaning during antenatal care from what she perceives. Relying as it does on aurally and not visually gathered information, her practice embodies and expresses the epistemic values on which her community has built a body of knowledge of the many complexities of sustaining human life in adverse socioeconomic and sociopolitical conditions.

But as with all epistemologies, we need to approach this acoustic epistemology (Feld 2012b) also as a modality of composing a world, an expression of the right and, indeed, the need to exercise ontological self-determination.[20] It is here that Pacific Afro-Colombian midwifery constitutes a politics in the face of dismissal and marginalization of alternative knowledge to that sanctioned by Western naturalism and by the Colombian state's biomedical institutions. Despite worldwide initiatives to promote mutual cooperation between midwifery and government public health systems, and despite statistics showing that midwifery could provide up to 90 percent of "prenatal" care, Colombia's health institutions have insisted on giving exclusive priority to biomedical care (Guarín and Molano 2016: 15). Midwifery practices in the region may appear as the introduction of alternative knowledge of the human body according to the naturalist formula "one nature, multiple epistemic perspectives on it," where their perspective ends up being repositioned to the status of cultural beliefs under the logic of cultural relativism. This is certainly the case in the recent governmental classification of Pacific Afro-Colombian midwifery as "Immaterial Cultural Patrimony," a status ASOPARUPA sought as a step toward a potential demarginalization of midwifery tout court.

Agreement over some fundamental aspects of the constitution of human physiology — that is, over the human body's unitary naturalistic character — does not merely reflect strategic allowances by midwives to biomedical naturalism. We all have a heart, they might say, and it is clear that they came to know this independently of the naturalist scheme, which cannot claim exclusive rights to truly know human physiological constitution. Likewise, they have come to agree about aurally detectable symptoms of a malfunctioning heart. Midwifery, in other words, has also developed notions of pathological

*Jairo Moreno*

normativity, and like most medical practices, it operates dynamically and in some sort of dialogue with other ways of knowing, especially in a governmental context in which midwives in Buenaventura, for example, today need to be vetted by the state's health system.

But these points of convergence really constitute a place from which it is the mode and manner of relation between one part (say, an agitated heartbeat in the fetus) and another (say, an acoustic milieu that may not be propitious to the well-being of the fetus) that demands that, if antenatal aurality constitutes a domain of amplified relationality, we consider its implications for the reality of bodies, an ontological question.[21] More pointedly, recall how the earliest auscultation during a pregnancy constitutes the instauration of aural relationality as what a life is, an ontological arrangement incomprehensible within naturalism's logic. The dynamics of comprehending those arrangements and negotiating them beyond but in relation to epistemology ask for a political understanding other than the liberal multicultural option of a politics of recognition of "difference"; than the constructivist critique that aims to debunk all ontological projects; or even than a critique of contemporary biopolitics that continues to rely on a naturalistic conception of human life and death to overcome "the mechanisms through which the basic biological features of the human species became the object of a political strategy, of a general strategy of power" (Foucault 2009: 1).

A politics that addresses things ontologically, as I suggest, begins with the admission that power differences, which define "politics" in much of our discourse and our expectations that addressing such differences might provoke change, exist in fundamental relation with the power of difference, which too often we link inextricably to our advocacy for the rights of others defined in the context of our reporting of their alternative ways of knowledge and culture.[22] Any thinking of the "alternative" (or alternativity) needs to go past the sheer fact of the difference of the Other (or alterity) so as to engage how the Other differs from itself—or, in our case, how the intensive differences emerge from and are constituted in aurality. And something like extensive differences emerges from the fact that there is a multiplicity of forms of existence for human death, to take a particularly pregnant example. As we have seen, the aural constitution of an infant's death is not the same as that of an adult, and the former, in turn, stands in relative contrast to a biomedical definition of a lack of vital signs, partly corresponding to that definition and partly contravening it. Even in the comparative case I made with deconstructionist perspectives on death (Derrida's life/death), which forcefully tries to cope with the shattering temporal implications of a death that always arrives but that one can never meet as one's "self," we could sense how such thinking can operate only within naturalist conventions as a metaphysical matter, the self-differing allowances of naturalism accountable through the categorical distinction between science and philosophy it enforces. For Mrs. Zambrano, even as she auscultates looking for signs of fetal well-being that would be fully in place at the local hospital, the coexistence of several forms of human existence take place both in concrete practices and practices of speculation. Of course, the "practical" and the "speculative" are heuristic categories resulting from my comparative exercise here. In reality, what I have called antenatal aurality convokes a "non-skeptical elicitation of manifold potential for how things could be," as Martin Holbraad, Axel Pedersen, and Viveiros de Castro (2014) propose. "How things could be": that is the question. It is not a matter of taking seriously what others believe the world to be, or what they take to be aurality's performativity in the composition of that world. It goes without saying that "taking seriously" marks a point of departure, not an end in itself. And as I have argued, the category of "belief" must be interrogated, if not altogether displaced, when things are thought ontologically—or, better, when the ontological gains full weight as a reasoned practice of and about the status of being.

But note that one could easily take this as an affirmation of "things as they are for others," not of "how things could be." The latter option emerges in the context of the comparative operation (Holbraad et al. 2014). The "could" in the phrase points to a potentiality drawn from engaging with what the thought of others affords us conceptually. Let me clarify. First of all, the engagement I refer to here occurs in these pages, as the impossible description of a network of aural relations produces an analytic of relationality writ large and using the borrowed language of various disciplines, both of which translate as much as they may disfigure but also open up some—and only some—form of access to other sides of things (Holbraad et al. 2014). This does not mean these sides can gain for oneself the reality it has for others, ontological multiplicity being the index of this impossibility. But this impossibility constitutes at the same time an index of a possibility—namely, that one comes to pass through what one studies, enacting through comparison the intensive difference that others already maintain in the living thought that is their practical metaphysics.

The expressions "practical metaphysics," "living thought," and "reasoned practice" all reflect my own effort (and that of others) to translate something that, according to the discussion of antenatal aurality in Pacific Afro-Colombia, just "is." But it is this "just is" that, as I have tried to show, holds

*Jairo Moreno*

remarkable conceptual richness and subtlety and reveals the forceful enactment of a wager on the very character of what "to be" means. Expressing and experiencing the fact of postmortem listening, for example, renders "to be" as a form of transit between domains (biological life and death, cosmological quasi-life and death, intensive and extensive differences) that is not simply the liberating "becoming" that we so often invoke as the default option to escape the constraints of that which "is." Antenatal aurality exists in a realm where "to be" and "to become" neither exist separately nor coexist (the hybrid option, which depends on a prior separation). Instead, in another mode of existence where these divisions do not operate, antenatal aurality, both as the specific labor of Mrs. Zambrano and many other midwives and as a general constellation of ideas and practices (the separation is my own), might reveal how, by comparison, from our own aurality, aurality could be.

Finally, a word about how the comparative exercise, with its axiomatic of ontology, might contribute to a cartography of sound studies. After all, the grand concept-building strengths of French philosophical anthropology and continental philosophy have the troubling effect of replicating the old colonial enterprise of subsuming the Other and its difference under schemes wholly other than their own, or of smuggling a Western transcendental subject of knowledge under the guise of—and, yes, the will-to-reason of—schematization and taxonomies that may well be another symptom of "Western" epistemology or of an analogist impulse, or an effect of naturalism's own binary structure of interior-exterior oppositions, as Descola's work has often critiqued for (Kohn 2015). Beyond the obvious fact that we would not be asking questions of sound were not disciplinary formations and their knowledge apparatuses already in place, there are a few things that merit consideration.

The first corresponds to an ethos of provincialization (Chakrabarty 2000) that shows "Western" thought to constitute one strain among several, with its set of parochial concerns, internal debates, and desire to gain meaning by understanding things in one particular way. But this strain of thought, as Dipesh Chakrabarty forcefully argues, cannot be critically comprehended either in splendid isolation or by simply refusing to engage it because of its pernicious, often devastating, effects on the world's peoples. Engaging Western thought means taking it seriously by learning its quirks and genealogies and studying its efforts as precisely that: efforts. Avoiding isolation means setting the scene for some sort of dialogue, a useful metaphor for comparative work, provided that it contains an agonistic element—thinking with and against is a premise of comparative work that introduces more than a modicum of misrecognition as it avoids turning comparison into a "*disposi-tif* of recognition: classification, predication, judgment, and representation" (Viveiros de Castro 2014: 43).[23] Here, for instance, I have brought Derrida to bear on the question of life and death not to show how his creative pro-posal remains caught in the web of naturalism's either-or structures—if anything, Derrida pushes hard against these structures. Instead, the notion of "life/death" reveals an effort to address and respond to the universality of human death by thinking its temporality (temporalities, more precisely) as an admittedly imperfect parallel to the way the Pacific Afro-Colombian people enact a related but dissimilar concern and do so without the limita-tions imposed by individual rationality. A first effect here is one of leveling the field insofar as these positions before "death" enter in a relation. This leveling field takes place within a politics of the otherwise—that is, with an eye on "how things could be," not within the fantasy of a cross-cultural con-nection. Furthermore, this is not to say that the two ontological positions are relative to the universal phenomenon of death. In positing the kinds of thinking of death that they do—where this thinking is equally and inextrica-bly a doing—the reality of death, its ontology, becomes something singular to each case while sharing an abiding concern with the shattering "reality" that humans do perish. It is a mistake to think that radical alterity implies that nothing is shared across ontological domains.

A second effect has to do precisely with what "perishing" means in each case. If for Derrida, to stick to his case, death exists as a constant future ante-rior ("I will have died") to a living human, in our case the dead live on for the living and for themselves as much as the living live their death for the dead and for themselves in a complex redoubling of relations sensorially enacted. The biological death that compels Derrida's elegant formulation also matters for Mrs. Zambrano, but it generates a set of relationships that take precedent over the *relata* (i.e., the living, the dead), on the one hand, and, on the other hand, repositions the place that biological (natural) death has in and for the community.

This repositioning of nature constitutes the second major issue. Antena-tal aural relationality does shift the ontological weight of what a life is to the relation itself, as I argue. But the inclusion of antenatal beings in social in-tercourse also implies an ontological presupposition other than naturalism's ascription of "nature" to that which is not social or constructed but given in advance by nature itself. Antenatal aurality operates with the presumption that a fetus shares with postnatal humans both their capacity for interiority and their external corporeality, and this is irreducible to cultural belief or

*Jairo Moreno*

social construction. As Eduardo Kohn (2015: 317) sums it up with reference to Descola, "Only with naturalism is 'nature' as an object external to our subjective selves conceivable."

In other words, without a binary opposition between nature and culture, analogism can be said to have no need to distinguish ontologically between them, because all things are same: if everything is nature and everything is culture (if everything shares in having an interiority and an exteriority), then nothing is culture and nothing is nature. But as my discussion of Mrs. Zambrano's clinical auscultation also shows, naturalism's "nature" plays a role in her understanding of antenatal life. It is just that it does not play *the* defining role. Given that there is only partial commensurability between naturalism's "nature" and midwives' "nature," the risk of subsumption of the latter by the former is simply neutralized—full commensurability being a requirement for subsumption to take place. Beyond this, it becomes clear that provincializing requires that the foundational ontological operation of naturalism's nature be remapped itself, and comparison offers one way to get this under way. Likewise, the work of decolonization could take seriously the root word for colony (the Latin *colere* 'to cultivate') to turn the soil in which the culture concept has so deeply taken root. Questioning the paradigmatic matrix of nature and culture, as many of the authors I cite have done recently, in turn compels a questioning and rearrangement of the universal and the particular, the objective and the subjective, the interior and the exterior, the continuous and the discontinuous, the given and the made, necessity and spontaneity, immanence and transcendence, body and spirit, humanity and divinity, and much more.[24] Aurality and its corollary "sound" have been taken to fall squarely into one side or the other in these syntagmatic pairs, but if the discussion of antenatal aurality here suggests anything, it is that any remapping in our sonic cartographies requires that determinations in this regard must be studied anew by the ontological arrangements that any collective may have made or may yet make.

*Notes*

1. The region covers the entire Pacific Coast of Colombia, and 83 percent of the population is Afro-descendent (Barbary and Urrea 2004: 73). It includes four Colombian provinces (*departamentos*): Chocó, in the north; Valle del Cauca and Cauca, in the middle; and Nariño, in the south. Coastal and richly fluvial, the region, although marked by specific variations across these four provinces, has long been considered by its communities a particular cultural and political *territory* in relation to the nation as a whole (Escobar 2008). Midwifery practices are consistent throughout the region, with

variations emerging in relation to particular places' having more or less contact with indigenous groups, as is the case of Chocó, or according to the use of various ingredients for cutting and suturing the umbilical cord (Portela Guarín and Molano 2016: 58).

2. I use the term "antenatal" instead of the more common "prenatal" in order to connote both a temporal precedent and a special sense of being "before" something else, or even "against" it. The implications of this "before" and "against" will become clear in the course of the chapter.

3. On "practical metaphysics," sometimes cast as "empirical metaphysics," see Latour 2005. Eduardo Viveiros de Castro (2014) refers to a "metaphysics of the others," and Marilyn Strathern (1990) writes of "Melanesian metaphysics."

4. "Para-human" denotes that which stands in some relation of exteriority to the human, broadly understood, as in the sense of "standing before" the human, as well as a sense of a transcendent temporality "prior to" the human but also coeval with it. Here, the latter denotation has to do with notions of a divine origin of humankind (see Surin 1998; Tomlinson 2016).

5. All references to Mrs. Zambrano are from an interview based on a questionnaire I developed that was administered by Janet Rojas, an anthropologist who has worked with women in the region for three decades, in August 2014, in Buenaventura, Colombia. For a beautiful image of Mrs. Zambrano at work, by Salvatore Ludicina Ramirez, please see https://www.elespectador.com/noticias/nacional/otra-forma-de-nacer -articulo-497377.

6. A total of 55 percent of the goods that come into Colombia make their way through Buenaventura. The port was privatized in 1993, and although it generates $4.2 million in taxes, a meager 0.17 percent is given to the city. Some 85 percent of the population live at or below poverty levels; 88.7 percent of the population is Afro-descendent. Only 71 percent of the city has access to potable water, although it often is available for only three hours a day. Criminal bands known as Bacrim, consisting of reinserted former paramilitary, control much of the city in ultraviolent turf wars for narcotic-trafficking routes. The homicide rate of 113 per 100,000 people is nowhere near the 33 per 100,000 rate in the rest of the country, which is besieged by armed conflict.

7. The association includes Afro-descendent male and indigenous female members, and it has been remarkably active in establishing dynamic exchanges with midwives in Africa (Senegal) and Central America (mainly Mayan women), holding an annual international meeting in Buenaventura.

8. "Conservar los saberes ancestrales culturales, y la práctica de la medicina tradicional con los conocimientos científicos, fomentando la unidad, solidaridad [e] interculturalidad, para estar siempre en una búsqueda permanente de saberes que nos ayuden a complementar y articular el sentido de pertenencia y de compromiso comunitario que nos anima a vivir en el pacífico colombiano ya que somos gestoras en apoyo a la vida" (Grupos Afroamericanos de Latinoamerica y el Caribe 2012).

9. This section develops material that is also treated in Moreno and Steingo 2018.

*Jairo Moreno*

10. Music plays a key role in the funerary rituals of children (see esp. Birenbaum Quintero 2018; Restrepo 2011).

11. In contrast, failure to observe proper funerary rites for dead adults (and children after age seven) renders them as visions that will continue to roam in places such as cemeteries or the jungle. There, the dead haunt the living, harassing them, and rendering such spaces inhospitable for them.

12. As he writes, this constitutes "an interminable experience [that] must remain such if one wants to think, to make come or to let come any event of decision or of responsibility" (Derrida 1993: 16).

13. I adopt the notion from what Viveiros de Castro describes as a *quase acontecimento*, referring to Amerindian indigenous peoples' conception of death as an ambiguous threshold between life and afterlife rather than simply an absolute limit between life and death: Eduardo Viveiros de Castro, "A morte come quase acontecimento," YouTube video, 2009, accessed January 20, 2017, https://www.youtube.com/watch?v=nz5ShgzmuW4.

14. I would be remiss not to note the wide range over which Derrida thought about death. For an excellent treatment, see Hägglund 2008.

15. As Eduardo Kohn (2015: 317) argues, "The assumption that others have dissimilar interiorities . . . orients . . . 'naturalism.' . . . [T]his is visible at a number of levels: [including] the individual (where solipsism and the problem of other minds are philosophical problems)."

16. David Toop (2010) takes the trope of mystery to be inextricably linked with our original acoustical dwelling in the amniotic sac. Critique of related literature appears in Moreno and Steingo 2018.

17. In Deleuze (2005), the plane of immanence is a plane of pure multiplicity in which everything that is exists and from which everything that comes to be issues. "A life," he writes (2005: 29), "is everywhere . . . : an immanent life carrying with it the events and singularities that are merely actualized in subjects and objects." Claire Colebrook (2006: 3) explains: "Deleuze refuses to see deviations, redundancies, destructions, cruelties, or contingency as accidents that befall or lie outside life; life and death were aspects of desire or the plane of immanence."

18. For succinct commentary on the ongoing dispute between Descola and Viveiros de Castro, see Latour 2009. See also Descola's and Viveiros de Castro's own critiques (Descola 2013; Viveiros de Castro 2014). Kohn (2015) offers a useful critical overview of Descola, Latour, and Viveiros de Castro.

19. Myth provides here what Viveiros de Castro (2014: 68) calls "a geometrical locus where the difference between points of view is at once annulled and exacerbated." In our case, the difference between listening perspectives (of divinities, of the unborn child, of the living) is at once annulled and exacerbated.

20. The reference is to Viveiros de Castro's oft-cited claim that anthropology's project is the "ontological self-determination of the world's peoples" (Holbraad et al.

2014). Steven Feld (2012b, xxxviii) approaches the ontological when he writes about "an acoustemological triangle connecting sound to ecology and cosmology."

21. As Annemarie Mol (2002) argues, attention to medical practice and health organizations shows how, beyond allowing a critique of authority, historical contingency of knowledge, and modes of governance, the reality of the human body is contested at an ontological level as to what that reality is.

22. My discussion of the political owes much to the argument presented in Holbraad et al. 2014.

23. According to Patrice Maniglier, for Viveiros de Castro, "A veritable anthropology . . . 'returns to us an image in which we are unrecognizable to ourselves,' since every experience of another culture offers us an occasion to engage in experimentation with our own — and far more than an imaginary variation, such a thing is the putting into variation of our imagination" (Viveiros de Castro 2014: 41, in Maniglier 2005: 773–74).

24. Ana María Ochoa Gautier (2016) makes an eloquent and forceful case in this regard in her expansive critique of ecomusicology, informed by the work of Viveiros de Castro.

## References

American Academy of Pediatrics, Committee on Environmental Health. 1997. "Noise: A Hazard for the Fetus and the Newborn." *Pediatrics* 100, no. 4: 724–27.

Arocha Rodríguez, Jaime. 1999. "Ombligados de Ananse: Hilos ancestrales y modernos en el Pacífico colombiano." Bogotá: Editorial Universidad Nacional de Colombia.

Barbary, Olivier, and Fernando Urrea, eds. 2004. *Gente negra en Colombia: Dinámicas sociopolíticas en Cali y el Pacífico.* Cali, Colombia: Editorial Lealon.

Birenbaum Quintero, Michael. 2018. *Rites, Rights, and Rhythms: A Genealogy of Musical Meaning in Colombia's Black Pacific.* New York: Oxford University Press.

Chakrabarty, Dipesh. 2000. *Provincializing Europe: Postcolonial Thought and Historical Difference.* Princeton, NJ: Princeton University Press.

Chion, Michel. 2016. *Sound: An Acoulogical Treatise,* trans. James A. Steintrager. Durham, NC: Duke University Press.

Colebrook, Claire. 2006. *Deleuze: A Guide for the Perplexed.* New York: Continuum.

Colombian Ministry of Culture. 2016. "Las parteras del Pacífico colombiano son patrimonio del país." Accessed January 19, 2017. http://www.mincultura.gov.co/prensa/noticias/Paginas/Las-parteras-del-pac%C3%ADfico-colombiano-son-patrimonio-del-pa%C3%ADs.aspx.

Deleuze, Gilles. 2005. *Pure Immanence: Essays on a Life,* introd. John Rajchman, trans. Anne Boyman. New York: Zone Books.

Derrida, Jacques. 1993. *Aporias,* trans. Charles Dutoit. Stanford, CA: Stanford University Press.

Descola, Philippe. 2013. *Beyond Nature and Culture,* trans. Janet Lloyd. Chicago: University of Chicago Press.

*Jairo Moreno*

Escobar, Arturo. 2008. *Territories of Difference: Place, Movements, Life, Redes.* Durham, NC: Duke University Press.

Feld, Steven. 2012a. *Jazz Cosmopolitanism in Accra: Five Musical Years in Ghana.* Durham, NC: Duke University Press.

Feld, Steven. 2012b. *Sound and Sentiment: Birds, Weeping, Poetics, and Song in Kaluli Expression*, 3rd ed. Durham, NC: Duke University Press.

Foucault, Michel. 2009. *Security, Territory, Population: Lectures at the Collège de France, 1977–78*, trans. Graham Burchell. London: Picador.

Grupos Afroamericanos de Latinoamerica y el Caribe. 2012. "24 años de ASOPARUPA (Buenaventura)," May 3. Accessed May 20, 2015. http://afrosenamerica.blogspot .com/2012/05/24-anos-de-asoparupa-buenaventura.html.

Hägglund, Martin. 2008. *Radical Atheism: Derrida and the Time of Life.* Stanford, CA: Stanford University Press.

Holbraad, Martin, Morten Axel Pedersen, and Eduardo Viveiros de Castro. 2014. "The Politics of Ontology: Anthropological Positions." *Cultural Anthropology*, January 13. Accessed May 1, 2017. https://culanth.org/fieldsights/462-the-politics-of-ontology -anthropological-positions.

Kohn, Eduardo. 2015. "Anthropology of Ontologies." *Annual Review of Anthropology*, no. 44: 311–27.

Latour, Bruno. 2005. *Reassembling the Social: An Introduction to Actor-Network-Theory.* New York: Oxford University Press.

Latour, Bruno. 2009. "Perspectivism: 'Type' or Bomb'?" *Anthropology Today* 25, no. 2: 1–2.

Lévi-Strauss, Claude. 1983. *The Raw and the Cooked: Mythologiques*, vol. 1, trans. John and Doreen Weightman. Chicago: University of Chicago Press.

Losonczy, Anne-Marie. 1989. "Del ombligo a la comunidad: Ritos de nacimiento en la cultura negra del litoral Pacífico Colombiano." *Revindi* 1: 49–54.

Manigliér, Patrice. 2005. "La parenté des autres (à propos de Maurice Godelier, Meta-morphoses de la parenté)." *Critique* 701 (October): 758–74.

Mol, Annemarie. 2002. *The Body Multiple: Ontology in Medical Practice.* Durham, NC: Duke University Press.

Moreno, Jairo, and Gavin Steingo. 2018. "The Alluring Object of the Heartbeat." In *Sound Objects*, ed. Ray Chow and James Steintrager, 167–84. Durham, NC: Duke University Press.

Nancy, Jean-Luc. 1993. *The Birth to Presence*, trans. Brian Holmes. Stanford, CA: Stanford University Press.

Nancy, Jean-Luc. 2007. *Listening*, trans. Charlotte Mandell. New York: Fordham University Press.

Navarro Valencia, Martha Cecilia. 2012. *Cuerpos afrocolombianos: Prácticas y representaciones sociales en torno a la maternidad, las uniones y la salud sexual y reproductiva en la costa Pacífica Colombiana.* Quito, Ecuador: Abya Yala.

O'Callaghan, Casey. 2009. "Sounds and Events." In *Sounds and Perception: New Philosophical Essays*, ed. Matthew Nudds and Casey O'Callaghan, 26–49. New York: Oxford University Press.

Ochoa Gautier, Ana María. 2016. "Acoustic Multinaturalism, the Value of Nature, and the Nature of Music in Ecomusicology." *boundary 2* 43, no. 1: 107–42.

Portela Guarín, Hugo, and Maria Elvira Molano. 2016. *Partería: Saber ancestral y práctica viva*. Bogotá: Banco de la República.

Povinelli, Elizabeth A. 2016. *Geontologies: A Requiem to Late Liberalism*. Durham, NC: Duke University Press.

Restrepo, Eduardo. 2002. "Comunidades negras del Pacifico Colombiano." *Guia del Museo de la Universidad de Antioquia*. Medellín, Colombia. Accessed January 15, 2017. https://www.academia.edu/2186910/Comunidades_negras_del_Pac%C3%ADfico _Colombiano.

Restrepo, Eduardo. 2011. "Representaciones y prácticas asociadas a la muerte en los ríos Satinga y Santianga, Pacífico Sur Colombiano." *Piedra de panduro*, revista de la Universidad del Valle, seccional Buga 8: 78–102.

Restrepo, Eduardo. 2013. *Etnización de la negridad: La invencion de las "comunidades negras" como grupo étnico en Colombia*. Popayán, Colombia: Editorial Universidad del Cauca.

Rice, Tom. 2013. *Hearing and the Hospital: Sound, Listening, Knowledge and Experience*. Canon Pyon, UK: Sean Kingston.

Sterne, Jonathan. 2001. "Mediate Auscultation, the Stethoscope, and the 'Autopsy of the Living': Medicine's Acoustic Culture." *Journal of Medical Humanities* 22, no. 2: 115–36.

Strathern, Marilyn. 1990. *The Gender of the Gift: Problems with Women and Problems with Society in Melanesia*. Berkeley: University of California Press.

Surin, Kenneth. 1998. "Liberation." In *Critical Terms for Religious Studies*, ed. Mark. C. Taylor, 173–85. Chicago: University of Chicago Press.

Tomlinson, Gary. 2016. "Sign, Affect, and Musicking before the Human." In *Econophonia: Music, Value, and the Forms of Life*, ed. Gavin Steingo and Jairo Moreno, *boundary 2* 43, no. 1: 143–72.

Toop, David. 2010. *Sinister Resonance: The Mediumship of the Listener*. New York: Bloomsbury Academic.

Viveiros de Castro, Eduardo. 2014. *Cannibal Metaphysics: For a Post-Structural Anthropology*, ed. and trans. Peter Skafish. Minneapolis, MN: Univocal.

# 5

# Loudness, Excess, Power

## A POLITICAL LIMINOLOGY OF A GLOBAL CITY

## OF THE SOUTH

*Michael Birenbaum Quintero*

.....

└                              ┘

This chapter examines people's practices of loudness in the Colombian city of Buenaventura to understand the imbrication of sounded practices in the workings of power in a "global city of the South" (Dawson and Edwards 2004). Buenaventura is a port city of some 300,000 inhabitants located on Colombia's Pacific coast, a rainforest region populated mostly by black Colombians and still in the process of integration into global modernity, in part through the influence of Buenaventura itself, the country's principal port and link to the economies of the circum-Pacific. Buenaventura is typical of Southern cities in that it is structured by densely reticulated regimes of politics and power: a system of governance belonging to one of Latin America's most continuously democratic states, a tradition of autonomous communities of the descendants of enslaved Africans (and some Amerindians), the technocratic management of the circuits of global capitalism, the biopolitical inculcation of liberal citizenship, a long colonialist (and now neoliberal) history of exclusion and abandonment, and a violent neocolonial integration. I follow practices of loudness to reveal how they exemplify local notions of political constituency, political action, experiences of sovereignty and abjection, and personhood itself. The local idiosyncrasies of these processes can ultimately help us rethink some of the assumptions undergirding theoretical studies of sound.

Daily life in Buenaventura is loud, by which I mean a few different things.[1] First, loudness is the cumulative effect of sound being everywhere. Trucks rattle, and motorcycles whinge; amplified music rattles zinc roofs and spills into the streets,[2] not only from dance halls and brothels, but also from homes, shops, cars, and buses; megaphone-distorted voices issue from the bicycles of street vendors and from storefront evangelical churches. To be more specific: reggaetón blasts from a minibus as a driver stops to hail a friend. In the back of the same bus, a boy plays music from his cellphone speakers. Hip-hop blares from a house in front of which a gaggle of teenage boys show off their dance moves to passing girls. Salsa echoes down narrow streets where neighbors sit in plastic chairs around a bottle of rum. Adolescent girls sing together with a record, at the top of their lungs. The ubiquity of loud sound has ramifications for spatial organization, constituting urban space as much as lines of sight and paths of transit. The aural experience of moving through the city is that of cross-fading from one sound system's radius of sonic immersion into another's. By "sound system," I mean in one sense the technological apparatus for amplifying sound,[3] but I also mean that each of these scenes—with its linked chains of sounding and listening agents and technologies of amplification—is a system of sorts. Thus, loudness is a social phenomenon, in its immersion of anyone occupying its sounded space, whether they are tarrying or passing by within earshot, in a field of exposure to a common sense world. Finally, although it seems obvious to say so, loudness is loud; that is to say, the mode in which one encounters many of these sound systems is that of a heightened experience of the body—rattling the sternum, hurting the ears, or having to shout to speak (cf. Heller 2015).

One place to begin to examine the meaning of loudness is with questions of power and of the political. Western political philosophy has long been haunted by the double nature of the political. There is a long-standing tradition of understanding and normalizing the political as a sphere of rational self-realization and mutual benefit, whether located in the formal political apparatus of the state or that of civil society. Achille Mbembe (2003: 12), drawing on Michel Foucault's work, has noted that this rationalist tradition effaces the more irrational workings of raw power, exemplified by the notion of the sovereign ruler, in which what is at stake is "the right to kill, to allow to live, or to expose to death." Scholars have depicted the relationship between rational and the irrational modes of politics in different ways. Some periodize the two, as in Foucault's account of the rule of the sovereign ruler (with his power to *make die and let live*) ceding to the rise of the apparatus of governmentality and the exercise of biopower that work, in turn, to *make live and let die* (Foucault [1976] 1990). For his part, Mbembe slices the pie geographically, offering that rational politics and modern biopower function in the

*Michael Birenbaum Quintero*

metropolises of the global North, while the colonies and postcolonies are the objects of the raw exercise of power because their supposed backwardness necessitates violence to undertake their civilization. Other scholars depict the two manifestations of the political as historically coeval and geographically coterminous, as in Domenico Losurdo's (2011) counterhistory of liberalism, which examines claims to freedom made by slaveholders; certainly, the oppression, expulsion, and even genocide of despised minorities internal to the polity by state actors casts a long shadow on the history of the nation-state, Northern and Southern alike.

Sound is intersubjective and thus broadly political, particularly in forms inflected by the exercise of power such as noise and loudness. Examinations of noise and loudness in sound studies therefore necessarily hark to some notion of the political. Part of my argument here is that much work in sound studies unreflexively assumes a rational, liberal, state-centered political frame for sounded practice. Thus, for example, the state is represented as having the power to produce, authorize, or condone loudness and to coerce silence (Bijsterveldt 2008; Braun 2011). Even in cases at the historical fringes of liberal modernity, in protomodern France (Corbin 1998) and the post-democratic War on Terror (Cusick 2006), descriptions of loudness focus on the imposition of loudness as a tool of power to shape or to pry apart human subjects. Loudness also becomes a tool of political contestation, when civil society usurps what we might call the state's legitimate monopoly on loudness to fill space with its own sound (see, e.g., Manabe 2013). The political model described in all of this work is essentially bipartite, with vested power opposed to either modern notions of civil society or older figures such as "community."[4]

Even studies that recognize the internal heterogeneity of civil society (see, e.g., Lee 1999) also recognize the role of the state as authorizing loudness or imposing silence, albeit less out of the brute exercise of power than the rational management of a chaotic aural public sphere (cf. Bijsterveldt 2008; Braun 2011).[5] As Jim Sykes (2015) has shown, this role is entirely in line with liberal notions of the state and with the rationalist notion of the political. In sound studies, this liberal genealogy underlies not only descriptions of the state, but also most constructions of listeners: citizen-individuals who withdraw themselves from the sounded public sphere when it impinges on them. This liberal listener appears in sound studies responding to public loudness or noise with retreat, aestheticization, or the technologically mediated individuation of the soundscape (see, e.g., Bull 2000; Keizer 2010; Schafer 1993). Here, the agonistic pair is not state and society but society and the individual, or public and private; the assumptions underlying their relationship are of the effective rule of law and the ability of private property to cloister the individual ear (thus enabling its self-actualization through attentive aesthetic listening) from public noise, a discrete and enforceable division of the public and private, and the sanctity of hermetic individuality.

The case studies I present from daily life in Buenaventura confound many of these assumptions, suggesting that a sound studies of the global South cannot conceive sonic actors or constituencies in the terms of the liberal political philosophies of the North.[6] First, agonistic understandings opposing society and the autonomous individual fail to account for the gamut of interpersonal entanglements (not to mention intrasubjective fragmentation) that characterize daily life in Buenaventura. So, too, do notions of the political that rely on counterposing either the state or a unitary sovereign against a monolithic notion of civil society, the public sphere, or "the people." The point is not that social mobilization is inconceivable for Buenaventurans; in fact, the city has a long, if always minoritarian, history of participation in the trade union movement, the Afro-Colombian social movement, the movement of displaced people and victims of violence, some student activism, and occasional highway blockages over services. Rather, I want to suggest that their political action is not limited to their potential impact on the sphere of political mobilizations, as some formulations aiming to describe postcolonial and global South contexts suggest (see, e.g., Chatterjee 2004; Holston 2009). I examine the ways in which a politics can be found in the everyday sphere, as practiced by that majority of the population that is not only marginalized from participation in the formal politics of the state but that also does not organize itself as a civil society mobilized against or seeking to enter that formal sphere.

Everyday politics and power struggles in Buenaventura, as revealed by practices of loudness, are more interpersonal and based on contingent and dynamic networks than on static constituencies. It is tempting to take the pessimistic view that people are exiled entirely from the sphere of the political by the brute violence of necropolitics. But I find that exclusion from formal politics, suspicion or agnosticism about social mobilization, and the precarity of everyday life are instead the conditions under which loudness practices can be understood as a bid for a certain kind of sovereignty, in that the simple act of psychic survival at a level beyond the barest of bare life (Agamben 1998) becomes in some ways political. This, of course, requires a broadening of notions of "the political" beyond the sphere of formal politics (Mouffe 2005), beyond teleological formulations of sociality as a standing reserve for social

*Michael Birenbaum Quintero*

movements (Holston 2009) or an alternative civil society (Chatterjee 2004), and perhaps even beyond the baseline notion of instrumental action toward a future. It is not that everyday sounded practices are apolitical, or, conversely, that they are reducible to politics, but that they are political in unrecognized ways (cf. Steingo 2016), at times even to their own practitioners.

This chapter arises from ethnographic experience in Buenaventura and its surroundings. The methodology that intervenes between fieldwork and theory is a study of the constitution of limits: a liminology (Moreno 2013). If loudness can be understood as the production of a particular brand of excess, a typology of limits proves useful. After all, where there is excess, there are limits being exceeded. Or not: some of the extreme volume I describe is not, in fact, excessive; describing it as such invokes limits that are not locally operative at all. In other cases, the limits being exceeded are fragmentations in the political community that are ingenuously glossed as "civil society" in the liberal political philosophies of the North. Tracing sonic excess, then, reveals limits that are either absent in or unforeseen by contemporary studies of sound.

### BUENAVENTURA, CITY AND PORT

Buenaventura is typical of cities in the global South, in which managerial or financial operations of a neoliberal bent and at a global scale coexist with contradictory regimes of everyday life at the lived scale, producing patterns of overlapping exclusion and inclusion in which the rational and irrational manifestations of the political are both temporally coeval and spatially intercalated (Davis 2004; Dawson and Edwards 2004; Sassen 2014). This overlap seems particularly relevant for Colombia, where one of the continent's most durable and stable democracies exists alongside a robust tradition of monstrous and lawless violence, a "state of exception" that is less the *suspension* of the rule of law than its Janus-faced corollary (Wade 2009).

The thinly populated rain forests of the Pacific region, of which Buenaventura is the largest city, are separated by the Andes from the rest of the country. The culture of the Pacific was forged by the descendants of enslaved people, whose society was founded less on enforced submission or armed revolt (although there was much of both) than simply waiting until the absentee slaveholders were otherwise occupied with their independence wars and slipping further into the boundless remoteness of the rain forest (Aprile-Gniset 2004). The response of the state to the ungovernability of peripheral territories such as the Pacific historically has been abandonment. During the twentieth century, however, with the completion of the Panama Canal and the rise of global demand for coffee (cultivated in the mountains adjoining the Pacific), Buenaventura was connected with the Colombian interior by railroad and began to claim importance.

Today, Buenaventura is Colombia's primary port, a hyperrationalized node in circuits of global capitalism connecting Colombia's national economy to the Panama Canal and Colombia's circum-Pacific trading partners. It is the primary point of entry and exit for international trade, handling 60 percent of the country's exports (Oxford Business Group 2014: 133). While Colombia also has ports on the Caribbean, the increasing importance of trade with Asia has given even more geostrategic importance to the port of Buenaventura. The port infrastructure is privately owned by multinational conglomerates (Oxford Business Group 2014: 133). But Buenaventura is also a city of 400,000 people,[7] literally walled off from the port enclave. It is a setting for bustling informal and often violent economies, residual rural cultural practices, and the city's mostly black, largely impoverished population, whose integration into rational modernity remains incomplete, incoherent, and unequal.

State management of Buenaventura aims to maximize the efficiency of the port by minimizing the drag constituted by the city. The privatization of the port, formerly one of the city's largest employers, in 1993 replaced unionized labor with subcontractors. A new elevated roadway sends tractor trailers above clogged city streets directly from the port to the highway. A $450 million infrastructure and logistics plan is dredging the bay to allow larger tankers to enter, and the highway that provides the only connection between Buenaventura and the rest of the country, a winding and congested two-lane road threatened by traffic, frequent landslides, and occasional blockages by disgruntled truckers' unions, is undergoing massive reconstruction, complete with multiple tunnels and bridges, to minimize shipping and travel time (Centro Nacional de Memoria Histórica 2015; Oxford Business Group 2014).

State management of the city, meanwhile, is typical of Foucauldian accounts of biopower as the power to make live and let die through the rational management of those endemic particularities of local populations represented as threatening to a normatively framed greater good: noncapitalist uses of time and a nonaccumulationist notion of the "good life"; collectivist rather than individualist spheres of agency; pathologized sexual practices and conceptions of family; the presence of mystical and divine forces in daily life; and so on.[8] To address these ostensible ills, the state and associated nongovernmental organizations have instituted family-planning cam-

*Michael Birenbaum Quintero*

paigns, training programs (for nonexistent jobs), entrepreneurship courses (for nonexistent markets), and cultural activities designed to keep "at-risk youth" out of trouble, organized around making live through the entrainment of capitalist uses of time and accumulationist notions of the good life. The city's entrepreneurial sector recently opened a luxury hotel, shopping mall, and movie theater, offering what in Buenaventura is a radically new conception of leisure. All of this constitutes the "make live" component of biopower. But the typically neoliberal approach to development, producing aspirations while neglecting employment, services, infrastructure, and education, combined with rank corruption by public officials, has rendered the make live component of biopower far less pervasive than "let die." Although the port generates some $2 billion per year in customs duties, the city lacks a decent hospital, sewers, and running water (even though the Pacific is one of the rainiest places in the world). Sixty-three percent of households in urban Buenaventura live in poverty, and the unemployment rate is more than five times the national average. Rural communities lack roads, law enforcement, healthcare, and reliable electricity (Centro Nacional de Memoria Histórica 2015: 58–62). Adding to this is that, since the 1990s, the Pacific has been disproportionately affected by the arrival in the area of Colombia's civil conflict—indeed, the city of Buenaventura is a major receiver of rural refugees. Buenaventura's homicide rate is among the highest in the country, although underreporting and new modalities such as the kidnapping and dismemberment of victims so that their scattered body parts cannot be assembled has made official homicide statistics unreliable (Centro Nacional de Memoria Histórica 2015). Governmentality? We should be so lucky.

Alongside abandonment, Colombia has a long tradition of integration by force. In the eighteenth century, the colonial authorities burned the settlements founded by runaways and scofflaws in the interior of Colombia's Caribbean region, resettling them in so-called reductions until, as one account put it, they were again "living by the sound of the Church bell" (Helg 2004: 34). More recently, in deference to this venerable tradition of leaving to raw power what the state is ill-equipped to handle, many of the functions of the state apparatus have been handed off to local operators unencumbered by any responsibility to the citizens. In this arrangement, the rational biopower that makes live and lets die yields to the necropolitical power of warring armed groups that instead make die and let, perhaps, live. Attempts to integrate the Pacific coast into global modernity, whether through the expansion of Buenaventura's port enclave into neighboring slums or the establishment of agro-industrial monocultures in the rain forest, have relied on armed paramilitary gangs associated with the state authorities and local captains of industry to intimidate, murder, and forcibly remove anyone who obstructs the march of capitalist expansion. In the 2000s, paramilitaries entered Buenaventura to combat leftist guerrilla insurgents, also massacring human rights activists and peasants who claimed land titles. They were able to expel the majority of guerrilla forces from the city after a paroxysm of violence in 2005 that included multiple prominent massacres and bombings. Since then, the factions have splintered, with bloody conflicts over urban space by paramilitary groups seeking strategic points for moving illicit commodities, providing the port industry with expansion space, reproducing themselves among the local populace, engaging in petty scams that usurp state-provided services, and protection rackets (Centro Nacional de Memoria Histórica 2015).

The "post-paramilitary" groups have retained areas of influence in various neighborhoods, where they are the de facto local authorities.[9] In a neighborhood I used to stay in, it is the local paramilitaries a family approaches to get electricity or a state health insurance card. They also provide a kind of policing, summarily executing local youths rumored to be thieves, indigents, drug users, and the mentally ill, as well as women deemed sexually disruptive. Their concerns are territorial, blocking people's passage through the so-called invisible borders between neighborhoods and forcibly displacing people from their homes, which is related to plans for the expansion of the port into the impoverished seaside neighborhoods. When local populations have resisted forced removal by the government, the pressure of the paramilitaries has been successful at uprooting them. Indeed, in 2014 alone, more than 25,000 people in Buenaventura (of 400,000) were forced from their homes (Centro Nacional de Memoria Histórica 2015: 235). Periodically, the federal authorities step in to militarize the city for a few months, subsequently withdrawing troops after a few high-profile arrests. Thus, daily experience in the city revolves around the constantly changing territorializations and deterritorializations that arise from a complex and overlapping map of necropolitical and governmental modes of power.

In broader theoretical terms, we might think of the people of Buenaventura as living under both biopolitical and necropolitical, rational and irrational, political regimes. Ostensibly citizens of a democratic republic, they are largely excluded in terms suggested by Partha Chatterjee (2004), who sees a discrepancy between, on the one hand, the theoretical category of the citizen and the popular sovereignty vested in it, and on the other, the

*Michael Birenbaum Quintero*

practical category of the population and the managerial policy invested in it. But Chatterjee's conception of the "managed population" does not quite obtain in Buenaventura, where the make live component of biopolitical governmentality is far weaker than the let die tradition of abandonment. Given that the people of Buenaventura are also subject to the necropolitical regime of the armed actors, they occupy the overlapping space of exclusion from formal politics, the weakness of the life-managing component of biopower, and abjection before the murderous component of necropolitics—that is, the articulation of let live with make die.

Thus, lived experience of Buenaventura is steeped in actual or possible violence.[10] One moves through threat and fear as one moves through the humidity of the air itself: heavily armed men in front of the corner store; the periodic destruction of electrical towers, taking out the electricity for days; tanks parked outside elementary schools; local women harassed by bored young men with assault rifles. People make complexly choreographed detours to avoid crossing invisible borders, or dangerous neighborhoods, or neighborhoods at war with their own, or the bad memories of the street corner where a loved one was killed. Death is gossip: when I ask, "What's Jaimer been up to?"[11] my friend's response is simply, "They killed him." Violence has entered the common repertoire for conflict resolution. Why did Quique leave for Bogotá? "A stupid drunken argument, but the guy turned out to be one of *those* guys, so I had to leave before breakfast." Yenny tells me, "This one paramilitary was bothering my sister all the time, like 'I'm going to rape you, I'm going to rape you,' so she told another *paraco*. They were in the same group, but he hated his guts anyhow, and he took care of him." Violence even becomes its own explanation and erects its own moral guidelines. The refrain, as among the townspeople in Patricia Madariaga's ethnography of violence in Colombia's Urabá region, was, "That's rotten that they killed him, but he must have done something" (Madariaga 2006: 85; cf. Rotker 2000). In other words, violence, the zero point of politics and the mechanism through which necropolitics functions, is not experienced as exceptional, unusual, or exterior to everyday ethical decisions or problem solving. It is not only exercised by discrete sovereigns *on* a population. It is also diffused through that population's own networks of sociality and use of space—and, as we will see, of sound. It is, in other words, a way in which power and abjection are experienced as part of people's lives, entangled with but not limited to either the state or the armed actors.

## BIOPOLITICAL CITIZENSHIP:
## MAKE SILENT AND LET RE/SOUND

The relationship between the port and the city is best exemplified by the new, sleekly glass-clad luxury hotel that stands directly between the port complex and the city's downtown. The hotel's glass, like the port complex's ID checkpoint, separates the visiting executives and officials staying at the hotel from the noisily pullulating humanity of the streets outside. Soundproof double-paned glass doors open to each room's balcony using a specially designed rotating airlock mechanism that preserves the preternatural hush of the carpeted and immoderately air-conditioned space within the hotel from the hurly-burly of the city outside.

Given the thought put into soundproofing the hotel, it is curious, then, that although the bustling downtown is characterized by the sounds of traffic; the calls of street vendors; and amplified music from stores and buses and the beer gardens of the nearby public park, it is actually not so loud compared with residential neighborhoods, especially from the hotel's upper floors. But, then, the loudness being guarded against is not quite the perception of high amplitude; instead, it is the intrusion of—the mere *perception* of—the city outside the hotel at all. If, as for Jean-Luc Nancy (2007), sound is a medium of shared exposure between people, then the sounds of the city at the hotel would be an entanglement with, even contamination by, others. The limit of the double-paned glass, then, is where the sonic cloister of the sovereign individual borders the unruly public sphere of the mass or the crowd. Thus the separation enacted by the limit of the hotel's soundproof glass is not only between two classes of people or the spaces allocated to each, but between two different modes of subjectivity through sound. We find on the inside of the glass a subject familiar in mainstream sound studies: the unwilling hearer, invested in a hermetic subjectivity that uses private property and the social segregation of space to shut out the sounded emanations of the denizens of the city outside.[12] The sonic sequestering of these individuated hearers from the broader populace has a direct corollary in the bio-power dictum of "make live and let die," in which to make live is to construct an inviolably individuated sound environment, and to let die is homologous with being left exposed to the contaminating sounds of the world. In other words, *to make silent and let sound.*

*Michael Birenbaum Quintero*

Beyond the inapplicability of the notion of the public sphere, the soundscape of Buenaventura's poor residential neighborhoods might instead be imagined as heir to a typically Latin American ungovernability-cum-abandonment. These neighborhoods are characterized by a proliferation of sound-producing devices. Powerful home stereos, televisions, cell phone speakers, and human bodies render sound, including other peoples' sound, a constant presence, exacerbated by the incomplete enclosure of flimsily constructed residences; the densely packed human collectivities; and the public occupation of streets, alleys, and other people's front yards—precisely the invasive human clamor that the hotel aims to shut out. But the loudness of the neighborhood—and it is at times deafening—does not necessarily mean excess. Sound does *not* irrupt the socio-spatial delineation of public from private, because the limit between the two cannot be assumed. Indeed, for local people, a quiet neighborhood, with no noise and with everyone cloistered in their homes, is described as *solo* (lonely) or even *muerto* (dead). Instead, neighborhoods stake claim to sonic liveliness, as in the neighborhood named Piedras Cantan, after the old salsa song about Puerto Rico as a place so joyous that even the "stones sing."[13]

In this sense, the sound systems create a form of civil society that is different from a Habermasian bourgeois public sphere. Rather than a setting in which neighborliness and good civic behavior require not imposing oneself, sonically or otherwise, on others; rather than a practice of sound based on social norms of respectful withdrawal, sound systems in Buenaventura are invested in engagement and entanglement across difference. Thus, we find a socio-sonic disposition in which subjects are expected not only to deliberately immerse themselves in the shared sound environments, but also to actively project their own sonority out into it.

The sound systems that produce this extreme volume are home stereos owned by working-class men. Take Ober, for example, who squints at me as I hand him a screwdriver, telling me about preamps and woofers and tweeters. Ober is a poor man, displaced by violence from his home city of Tumaco. His earnings on construction jobs seem to go more toward his large stereo than toward the ramshackle house where he lives with his wife and daughters. "It's good to have a good loud stereo and good music to play on it," he says. "Everybody knows me around here. I've got the good melodies." The sound system, and the specialized technical knowledge it requires, is a means for Ober to project his own musical personality outward into local space (the louder the sound, the wider its spatial radius); to make himself significant, despite his marginality in larger society, as a sonic presence in the neighborhood; and to build a social network for himself that revolves around his stereo and the cold beer his wife sells from a wheezing refrigerator to the neighborhood people who gather there. "This machine is a public service," he shouts at me over the music, grinning. "If I stop playing music on a Sunday afternoon, people say, 'Hey, put the music back on!'"

But the community composed of this sonic space and the networks that wend through it is not evenly distributed; nor is it uncontested. It is, like any community, not given but finite and contingent—what Jean-Luc Nancy (1991) calls *desouvrée* (inoperative, unworked, or worked loose of itself). As such, it is constantly reshuffling its own limits, internal *and* external. Internally, it is shot through with the reticular networks of affinities and conflicts of which the micropolitics of the everyday are made. As much as distinctions between public and private are not always already operative, antagonistic intracommunity fractures can appear at any moment; they can be gendered, generational, racial, or interfamilial but most often are affinal, especially around the reshuffling of networks of intimacy that arise from sexual relations. These fracture lines are frequently traversed by sonic excess, such as when a betrayed spouse starts shouting curse words at the home of the lover's family at 3 AM. Or when a jilted lover publicly blasts a song about love's perfidy, on repeat, for several hours. Or when local evangelicals set up megaphones outside in the hangover-induced relative silence of a Sunday morning. Or when the paramilitaries decide to make their presence felt. The external limits of the loud neighborhood do not delineate a resistant, romantic aural *Gemeinschaft* disenchanted by a bourgeois/liberal sound ideology: the neighborhood dwellers do not care whether rich people in a hotel can or cannot sleep. Thus, the city borders the port through what Jairo Moreno (2013) calls a differential model of two systems that do not quite separate, or negate, or encompass one another.

NECROPOLITICAL ABJECTION:
MAKE RE/SOUND AND LET BE SILENT

It is necropolitics that structures the organization of urban space. Like the constantly reshuffling social map of Nancy's desouvrée community, the urban spatial map becomes a shifting patchwork of territories occupied by different armed actors, delimited by the so-called invisible borders, the

*Michael Birenbaum Quintero*

crossing of which is punished by execution. For this and other offenses, the paramilitaries have begun taking their victims to houses abandoned by fleeing families, where the victims, still alive, are hacked apart with axes and machetes. The dismembered body parts are wrapped in plastic bags, stored in refrigerators, and scattered; some are tossed into the sea, and some are placed as a warning in public places such as soccer fields. These "cutting houses" (*casas de pique*) are well known in their neighborhoods; the neighbors see the victims taken in full daylight, and their screams and pleas are quite audible, unless someone in the cutting house starts to blast the stereo.

The stereo is *not* a means to dissemble what is happening in the house by covering the victims' screams: everyone already knows what is happening, and, at any rate, witnesses describe the stereo as being turned on after, not before, the screaming starts—or it is not turned on at all.[14] The sound emanating from the cutting house limns the frontiers of public and private within the neighborhood much like the jilted lover's amplified dramatizations of heartache or social outrage mentioned earlier—by broadcasting what happens in the interior of one house into a wider space, an effect facilitated by paramilitaries' explicit or tacit prohibition on neighbors turning up the volume of their own stereos (Toro 2016). The sounds of the cutting house fill the space of the neighborhood and may overflow the limits of the invisible frontiers, warning neighboring rival territories of the spatial injunction.

Amplified music in the cutting house also exceeds other, internal limits—limits that are within the subjectivity of the victimizers themselves. One important reason for dismembering victims is pedagogical. Many of those who participate in dismemberment are children and teenagers (Centro Nacional de Memoria Histórica 2015: 286–90; cf. War Child et al. 2013). The paramilitaries train them to accustom themselves to the act of killing by having them do it in a particularly horrible and spectacular way that is also a social ritual of coparticipation. Here, music's temporal synchronization, known for lightening the loads of laborers from the factory floor to the chain gang, conspires with its extreme volume to help them carry out their gruesome task, not by blotting out the victims' screams, but, through the simultaneous heightening and demolishing of the senses that Michael Heller (2015: 44–46) describes as "listener collapse," by providing for a kind of disjuncture between the visual and the aural planes. This makes possible at least a partial disassociation from a visual plane that might be reconfigured into an almost cinematographic form of spectatorship, or to short-circuit what Don Ihde (2007: 37–38, 139), riffing on Edmund Husserl, calls the "mineness" of experience, by which the self is divined from the experience of the world.

Gold mining has become an important if only semilegal economic activity in the Pacific, sprouting from the cracks between a state apparatus pushing economic development but unable to regulate it and local armed actors who have found it inordinately profitable. The region's mining camps are gravelly, mercury-stained scars carved out of the rain forest. Aside from the mine itself, the camp, or a nearby town, always has the twinned pair of the gold buyer and the cantina, the mouth and asshole of mercenary capitalism in the Pacific. There, the white buyer accepts the gold dust wrapped in a sweaty ball of notebook paper the miner pulls from his rubber boot;[15] the miner presses the wads of discolored bills he receives into the hand of the cantina owner to buy another bottle of rum or, perhaps, into the hand of the blank-faced woman he grinds his bodily drunkenly against to the sound of the sound system. The mood is sullen and aggressive, not only among the men to whom the finder of the gold distributes liquor and provides music, but even by the miner toward himself, as drunkenness and the earsplitting loudness of the music become a kind of pleasurably agonizing pushing of the body to the limits of its physical endurance—again, the limits internal to the self. This loudness is the end product of a transaction in which the backbreaking work of the mine is translated not into the workers' accumulation of capital or economic betterment or even conspicuous consumption. Not only the ritual distribution of liquor toward one's fellows to bask in the prestige of their respect but also the sheer quantity of liquor, the volume of the sound system, and the apocalyptic scale of the drunkenness are something like a spectacle of their own destruction (as well as the destruction of the gold that purchased them and even of the worker's body that extracted the gold), much as the destruction of the potlatch works simply to show that so much can be destroyed (Bataille 1993; Mauss 1967). The limit transgressed here is value. This excess is unproductive, even pointless outside its own deliberate and wasteful destruction. This is the irony of the centrality of this waste in the Pacific gold mine—a mercenary hyper-capitalism that, for the worker, ultimately results in the production and subsequent destruction of excess quite alien to standard microeconomics. It is not accidental that loudness is an essential component of the ritual of excess. As Richard Leppert (1993: 26), writes about music in a very different time and place, "Since music itself is 'simply air,' what it encodes of power is the power over time. For some, the world can be made to stop while the air is filled with music's emptiness.

*Michael Birenbaum Quintero*

Music's power, according to this formulation, is its emptiness, which constitutes itself ironically as a sign of fullness and excess."

Back in Buenaventura city, the groaning bass of Mexican-style tears-in-your-beer music, pushed to inhuman volume by huge speakers, sets the walls of precarious wooden houses rattling while inside, mother, father, children, cousins, grandmother, and friends sing along together, with fierce joyfulness, at larynx-shredding volume, proclaiming in the space between the lines "Oye esto!" (Hear this!) or "Ay, jueputa!" (Ah, son of a bitch!), investing even the most banal lyrics of sentimental romance with a ferociously pleasurable intensity of feeling that transcends their literal meaning and seems permeated with those words that cannot be spoken. One night, joking, I ask Doña Sara after her spirited singing of one particularly maudlin song about an improbable love triangle, "Did that ever happen to you?" She looks at me with a faraway expression and eyes bleary with drink. "No, *m'ijo*, but the things that have. . . ." I know those things include poverty, divorce, a car accident, and a stint of ersatz prostitution in a Putumayo boomtown, but when her eyes well up at the end of the night, it is for her nephew that she calls out—her favorite nephew, whom she raised, who was killed, and whose body, reduced by dismemberment, she reclaimed for burial. She has told the story before, but tonight she just keeps singing, at the top of her lungs, on the verge of weeping.

The starting point for political change would seem to be language. But how can one speak of these things in Buenaventura against the insufficiency of language itself, the lack of a socially recognizable discourse of mourning, fear of reprisal (Madariaga 2006)? Not speaking is an imposed condition of political abjection, but it is also a tactic that allows those subject to violence to have some semblance of "normal" life. As the anthropologist Robin Sheriff (2001: 75) notes, "The paradox of silence resides precisely in the fact that it implies and objectively constitutes a kind of political accommodation to oppression at the same time that it allows people . . . to let it go, to forget it, to at least partially contain the wounds of victimization and carve out a world in which to live with dignity and laughter."

But, of course, in Doña Sara's house, or Ober's, or in the mining camp, or even in the cutting house, not speaking is not silent at all but deafeningly loud. In Doña Sara's house, singing through loudness is less narrative than transductive—a transfer of energy between fields of experience (Helmreich 2007), an enactment of the very experience of being swollen with emotional experience begging for release and of butting up against the limits not only of language but also of the voice. One is lacerated by sound as the speakers strain against the sheer volume that gushes out of them, singing with that sound at a volume that pushes the very breaking point of the larynx, with the physicality of pushing through the soreness of the throat the only perceptible element of a voice that, although one's own, cannot be picked out from the mingled boom of the recording and the voices of the others who, arms over one another's shoulders, sing, too, with a fierce and odds-defying joyfulness, a kind of asymptotic and unachievable pushing out that is constituted precisely by the limits of one's throat, of the stereo's speakers, of one's ability to hear oneself over the music, of the speakability of experience.

## WHAT IS THE POLITICAL OF SOUND?

Perhaps even more essential than language in most understandings of politics is as an orientation toward temporality—specifically, toward futurity. The Colombian state keeps the focus on a future, after the cease-fire with the Fuerzas Armadas Revolucionarias de Colombia (Revolutionary Armed Forces of Colombia), after the maximum efficiency of the port is achieved. In the city, however, daily experience offers no postconflict moment, no healing reintegration of the moral order, no material advancement. Under the specter of the necropolitical, time is experienced as an undifferentiated present that is essentially a permanent impasse. Locked into a never-ending holding pattern, constantly oriented to the ever changing vicissitudes of survival in the present, the future recedes beyond the horizon. The good life is not deferred—deferred to when?—it is necessarily experienced in the present, which is precisely where Georges Bataille (1967, 1993; see Esposito 2015) locates a limited kind of sovereignty in a fleeting and sacred present moment in which the subject for once is not oriented fearfully toward the future or its own death. Thus, the mechanisms of sovereignty in the present are also the mechanisms that ensure abjection. A focus on the present is also a disavowal of the future—that is, it normalizes the present regime of violence and abrogates the possibility of change in the future. This is both the strength and the weakness of practices such as loudness in a context in which politics is so profoundly constrained.

Practices of loudness—in the neighborhood, in the mine, in the home, in the cutting house—do not make violence speakable and therefore cannot serve as a means of reclaiming justice or publicly repudiating violence. They are, at best an exercise of "lateral agency" (Berlant 2007), a stopgap measure, although the continuing intractability of violence has made them a permanent one. But as a deeply embodied and heightened experience of present-tense

Michael Birenbaum Quintero

temporality, as a projection outward into urban space of a sound system that puts into mutual exposure a community of people placed at the limits of speech, the exterior of political participation, the insufficiency of gold, and the experience of death, loudness is not nothing, either.

### Notes

I am grateful to Gavin Steingo and Jim Sykes for inviting me to be a part of this project. I appreciate their helpful comments and those of audiences in the United States and Colombia who have heard versions of this project. I hope that the people of the neighborhoods in which I have lived and spent time in Buenaventura — Carlos Holmes Trujillo, El Progreso, Las Palmas, el Lleras, Camilo Torres, Punta del Este, and Centro — over the past twelve years can recognize something of their lives in this chapter.

1. Here, as throughout this chapter, I rely on experiences I have had in Buenaventura. Since 2005, I have returned to the city almost every year for a week or so to several months. Some of these visits have been explicitly dedicated to ethnographic fieldwork, but the distinction between research and the rest of life is blurry, because my wife and her family are from the city, and my son spent the first six years of his life there. The people whom I describe in this chapter are real people I know, with their names changed. At times, as in the paragraph to which this endnote is attached, I refer to recurring scenes of daily life with recourse to the indefinite pronoun because these tableaus are so common.

2. I am not concerned here with music as music, or with rehearsing the familiar ethnomusicological trope of music as metonymic of a place or social group. Music is important here as one particularly common kind of hyperamplified ludic sound in daily life.

3. I am, of course, also appropriating the Jamaican term for a large amplification system, although I use the term far outside the culturally specific Jamaican sense.

4. For Cusick (2006), the figuration is a bit different but structurally similar: power and personhood.

5. Supposedly. In many contexts, some participants in civil society have more right to sonic control than others (Sakakeeny 2010).

6. Indeed, this may be the case for the global North, as well, whether because the North is increasingly becoming precarious in ways already seen in the South (Comaroff and Comaroff 2012), or because entanglement remains despite assumptions of individuation as the natural state of the modern subject (Biddle 2007).

7. This figure, like some of the statistics cited in this chapter, includes both the city of Buenaventura and the larger municipality, which includes a number of rural communities spread over a wide geographical area.

8. It is worth noting that sound generally and loudness specifically play roles in the traditional cosmology: the role of the bombo drum is frightening away the *tunda* at children's funerals (Whitten [1974] 1994: 100), for example. See also Birenbaum Quintero 2010; Losonczy 2006.

9. A number of terms have proliferated to describe these groups — "neo-paramilitary," "criminal band," "illegal armed actor" — since the umbrella paramilitary organization

signed an accord with the state to demobilize in 2005. I generally use "paramilitary." The rise of these terms is intended, in part, to obscure the historical relationship of these groups with the state and business interests. "Paramilitary" and the somewhat dismissive slang term *paraco* are used most often in Buenaventura.

10. Violence is a central component of daily life in Buenaventura, and particularly of daily negotiations with power, which ultimately is the theme of this chapter. For many readers situated in the global North, merely describing violence in the explicit terms in which it is experienced may read like a kind of complicity—titillating the reader with morbid details, reducing daily life to violence, or, worse yet, presenting ethnography as a kind of extreme sport. My approach to writing about violence is indebted to a long tradition in Colombia itself of grappling with the ramifications of writing violence in a nation both wearied by violence and in which violence is nothing less than historically foundational (Ospina 2013). The dilemma is that, while describing violence can be read as sensationalistic, ignoring it is tantamount to authorizing it. The best I can do is to represent violence in sober, unsensational terms as part of, but by no means all of, people's lived experience. While this is no guarantee that my account will not be read sensationally, I might also note that the reader, too, has an obligation to engage responsibly with other people's experiences of violence, neither wallowing in them nor pretending they do not exist. The South African novelist J. M. Coetzee (1986: n.p.) makes a useful summary of the stakes here: "When the choice is no longer limited to either looking on in horrified fascination as the blows fall or turning one's eyes away, then the novel can once again take as its province the whole of life, and even the torture chamber can be accorded a place in the design." It is with the idea that my analysis can offer a take on violence in Buenaventura that is not limited to titillation, and the assumption that the reader is capable of looking on with something more critical than horrified fascination, that I can represent violence's place within the whole of life in this chapter. I thank Gavin Steingo for the Coetzee article.

11. As noted earlier, I use pseudonyms throughout the chapter.

12. Writing from the historical vantage point of the United States in July 2016, this sonic *cordon sanitaire* has clear resonances with Donald Trump's wall, Barack Obama's massive deportations of undocumented immigrants, Brexit, police harassment of black and brown people in segregated and gentrifying cities, proposed bans on Muslim immigration and, during the Ebola epidemic, on West Africans, and the migrant crisis in Europe. The fear of contamination and the perceived right to avoid contact with difference in a world in which people are already entangled on the capillary level have led to calls for increasingly draconian means of social prophylaxis. Although clearly only one dimension in which these dynamics play out, sound is nonetheless clearly important, as Matt Sakakeeny's (2010) work on New Orleans noise ordinances and the killing of the black teenager Jordan Davis at a gas station in Florida over Davis's loud music make clear.

13. Lebrón Brothers, "Las piedras cantan," on *Tenth Anniversary*, audio LP, Cotique CS-1093, 1977.

*Michael Birenbaum Quintero*

14. There are varied accounts of the cutting houses. My gut sense, which would require more investigation to confirm, is that in the seaside neighborhoods the cutting houses do not usually play music, while those of the inland neighborhoods of Comuna 12 do.

15. "The miner," like the mine I describe here, is a recurring, rather than specific, personage.

## References

Agamben, Giorgio. 1998. *Homo Sacer: Sovereign Power and Bare Life*. Stanford, CA: Stanford University Press.

Aprile-Gniset, Jacques. 2004. "Apuntes sobre el proceso de poblamiento del Pacífico." In *Panorámica afrocolombiana: Estudios sociales en el Pacífico*, ed. Mauricio Pardo Rojas, Claudia Mosquera, María Clemencia Ramírez, 269–90. Bogotá: Instituto Colombiano de Antropología e Historia.

Bataille, Georges. 1967. *La limite de l'utile: Oeuvres complètes*, vol. 7. Paris: Gallimard.

Bataille, Georges. 1993. *The Accursed Share: An Essay on General Economy*, vols. 2–3. New York: Zone.

Berlant, Lauren. 2007. "Slow Death (Sovereignty, Obesity, Lateral Agency)." *Critical Inquiry* 33: 754–80.

Biddle, Ian. 2007. "Love Thy Neighbour? The Political Economy of Musical Neighbours." *Radical Musicology* 2. Accessed June 27, 2018. http://www.radical-musicology.org.uk/2007/Biddle.htm.

Bijsterveldt, Karen. 2008. *Mechanical Sound: Technology, Culture, and Public Problems of Noise in the Twentieth Century*. Cambridge, MA: MIT Press.

Birenbaum Quintero, Michael. 2010. "Las poéticas sonoras del Pacífico Sur." In *Músicas y prácticas sonoras en el Pacífico afrocolombiano*, ed. Juan Sebastián Ochoa Escobar, Carolina Santamaría Delgado, and Manuel Sevilla Peñuela, 205–36. Bogotá: Universidad Javeriana.

Birenbaum Quintero, Michael. 2018. *Rites, Rights, and Rhythm: A Genealogy of Musical Meaning in Colombia's Black Pacific*. New York: Oxford University Press.

Braun, Hans-Joachim. 2011. "Turning a Deaf Ear? Industrial Noise and Noise Control in Germany since the 1920s." In *The Oxford Handbook of Sound Studies*, ed. Trevor Pinch and Karen Bijsterveldt, 58–78. New York: Oxford University Press.

Bull, Michael. 2000. *Sounding Out the City: Personal Stereos and the Management of Everyday Life*. Oxford: Berg.

Centro Nacional de Memoria Histórica. 2015. *Buenaventura: Puerto sin comunidad*. Bogotá: Centro Nacional de Memoria Histórica.

Chatterjee, Partha. 2004. *The Politics of the Governed: Reflections on Popular Politics in Most of the World*. New York: Columbia University Press.

Coetzee, J. M. 1986. "Into the Dark Chamber: The Novelist and South Africa." *New York Times*, January 12. Accessed July 22, 2016. https://www.nytimes.com/books/97/11/02/home/coetzee-chamber.html.

Comaroff, Jean, and John L. Comaroff. 2012. *Theory from the South: Or, How Euro-America Is Evolving toward Africa*. New York: Routledge.

Corbin, Alain. 1998. *Village Bells: Sound and Meaning in the Nineteenth-Century French Countryside*. New York: Columbia University Press.

Cusick, Suzanne G. 2006. "Music as Torture/Music as Weapon." *Revisit TRANS* 8. Accessed July 31, 2016. http://www.sibetrans.com/trans/articulo/152/music-as-torture-music-as-weapon.

Davis, Mike. 2004. "The Urbanization of Empire: Megacities and the Laws of Chaos." *Social Text* 22, no. 4: 9–15.

Dawson, Ashley, and Brent Hayes Edwards. 2004. "Introduction: Global Cities of the South." *Social Text* 22, no. 4: 1–7.

Esposito, Roberto. 2015. *Categories of the Impolitical*, trans. Connal Parsley. New York: Fordham University Press.

Foucault, Michel. [1976] 1990. *The History of Sexuality, Volume 1: An Introduction*. New York: Random House.

Helg, Aline. 2004. *Liberty and Equality in Caribbean Colombia, 1770–1835*. Chapel Hill: University of North Carolina Press.

Heller, Michael. 2015. "Between Silence and Pain: Loudness and the Affective Encounter." *Sound Studies* 1: 40–58.

Helmreich, Stefan. 2007. "An Anthropologist Underwater: Immersive Soundscapes, Submarine Cyborgs, and Transductive Ethnography." *American Ethnologist* 34, no. 4: 621–41.

Holston, James. 2009. "Insurgent Citizenship in an Era of Global Urban Peripheries." *City and Society* 21, no. 2: 245–67.

Ihde, Don. 2007. *Listening and Voice: Phenomenologies of Sound*, 2d ed. Albany: State University of New York Press.

Keizer, Garrett. 2010. *The Unwanted Sound of Everything We Want: A Book about Noise*. New York: Public Affairs.

Lee, Tong Soon. 1999. "Technology and the Production of Islamic Space: The Call to Prayer in Singapore." *Ethnomusicology* 43, no. 1: 86–100.

Leppert, Richard. 1993. *The Sight of Sound: Music, Representation, and the History of the Body*. Berkeley: University of California Press.

Losonczy, Anne-Marie. 2006. *La trama interétnica: Ritual, sociedad y figuras de intercambio entre los grupos negros y emberá del Chocó*. Bogotá: Instituto Colombiano de Antropología e Historia and Instituto Francés de Estudios Andinos.

Losurdo, Domenico. 2011. *Liberalismo: A Counter-History*. Brooklyn, NY: Verso.

Manabe, Noriko. 2013. "Music in Japanese Antinuclear Demonstrations: The Evolution of a Contentious Performance Model." *Asia-Pacific Journal* 11, no. 42. Accessed July 30, 2016. http://apjjf.org/2013/11/42/Noriko-Manabe/4015/article.html.

Mauss, Marcel. 1967. *The Gift: Forms and Functions of Exchange in Archaic Societies*. New York: Norton Library.

Madariaga, Patricia. 2006. *Matan y matan y uno sigue ahí: Control paramilitar y vida cotidiana en un pueblo de Urabá*. Bogotá: Universidad de los Andes.

Mbembe, Achille. 2003. "Necropolitics." *Public Culture* 15, no. 1: 11–40.

Moreno, Jairo. 2013. "On the Ethics of the Unspeakable." In *Speaking of Music: Addressing the Sonorous*, ed. Keith Chapin and Andrew H. Clark, 212–41. New York: Fordham University Press.

Mouffe, Chantal. 2005. *On the Political*. New York: Routledge.

Nancy, Jean-Luc. 1991. *The Inoperative Community*, trans. Peter Connor, Lisa Garbus, Michael Holland, and Simona Sawhney. Minneapolis: University of Minnesota Press.

Nancy, Jean-Luc. 2007. *Listening*, trans. Charlotte Mandell. New York: Fordham University Press.

Ospina, William. 2013. *Pa que se acabe la vaina*. Bogotá: Planeta.

Oxford Business Group. 2014. *The Report: Colombia 2014*. London: Oxford Business Group.

Rotker, Susana. 2000. *Cuidadanías del miedo*. Caracas: Editorial Nueva Sociedad.

Sakakeeny, Matt. 2010. "'Under the Bridge': An Orientation to Soundscapes in New Orleans." *Ethnomusicology* 54, no. 1: 1–27.

Sassen, Saskia. 2014. *Expulsions: Brutality and Complexity in the Global Economy*. Cambridge, MA: Harvard University Press.

Schafer, R. Murray. 1993. *The Soundscape: Our Sonic Environment and the Tuning of the World*. Rochester, VT: Destiny.

Sheriff, Robin. 2001. *Dreaming Equality: Color, Race, and Racism in Urban Brazil*. New Brunswick, NJ: Rutgers University Press.

Steingo, Gavin. 2016. *Kwaito's Promise: Music and the Aesthetics of Freedom in South Africa*. Chicago: University of Chicago Press.

Sykes, Jim. 2015. "Sound Studies, Religion and Urban Space: Tamil Music and the Ethical Life in Singapore." *Ethnomusicology Forum* 24, no. 3: 380–413.

Toro, Juan José. 2016. "Así resiste Puente Nayero, un oasis en medio de la guerra en Buenaventura." *Pacifista!* Accessed June 27, 2018. http://pacifista.co/asi-resiste -puente-nayero-un-oasis-en-medio-de-la-guerra-en-buenaventura.

Wade, Peter. 2009. "Inside Out: The Changing Dynamics of Inclusion and Exclusion for Ethnic Citizens in Latin America." Centre for Research on Socio-Cultural Change, Economic and Social Research Council, Government and Freedom: History and Prospects Seminar Series, Seminar 4: Liberal Government and Its Outsides. *Open University*, June 12. Accessed June 27, 2018. https://www.academia .edu/7476880/Inside_out_the_changing_dynamics_of_inclusion_and_exclusion _for_ethnic_citizens_in_Latin_America.

War Child, Fundescodes, Servicio Jesuita a Refugiados, and Coalición contra la Vinculación de Niños, Niñas y Jóvenes al Conflicto Armado en Colombia. 2013. *Niños, niñas y adolescentes en busca de la Buena Ventura*. Bogotá.

Whitten, Norman E., Jr. [1974] 1994. *Black Frontiersmen: Afro-Hispanic Culture of Ecuador and Colombia*. Prospect Heights, IL: Waveland.

# 6

# The Spoiled and the Salvaged

## MODULATIONS OF AUDITORY VALUE

## IN BANGALORE AND BANGKOK

*Michele Friedner and*

*Benjamin Tausig*

.....

### INTRODUCTION: SPOILED AND SALVAGED

In this chapter, we consider how the categories of "hearing" and "sound" modulate between the values of the spoiled and the salvaged.[1] We investigate the binary opposition between what is devalued and what might (yet) provide value, and we think through the ways that sound, hearing, and deafness afford value. We are also called to examine how actors modulate among other seemingly categorical binaries: the "global South" and the "global North," deaf and hearing, and ability and disability. Our approach is collaborative and ethnographic. As a medical anthropologist (Friedner) and an ethnomusicologist (Tausig), we present two ethnographic scenes from our research sites in a dialogue that attends to the creation of value in the interplay of social and economic processes as they relate to hearing and deafness.

In considering sign-language-using deaf young adults in Bangalore, India (Friedner), and a deaf car stereo installer in Bangkok, Thailand (Tausig), we explore how the categories of the global North and the global South are constructed in active and collaborative engagement with each other. Attending to the experiences of deaf people living in the global South provides a potent means of tracking how the global North—through discourses, political and economic power, and international treaties and institutions—is both mobilized and extended, and how categories are made universal or not. In our two urban sites, Bangalore and Bangkok, inequality and economic mobility coexist, and the local and global are intertwined. How do hearing and deafness afford value in these spaces? In approaching this question, we consider, especially through the work of David Graeber (2013), how value creates social, political, and economic—as well as (non)auditory—worlds.

To these ends, we think athwart biology in our understanding of hearing (Helmreich 2011).[2] In an implicit critique of hearing's biological basis, Hillel Schwartz (2011: 22) stresses that many scholars too readily assume "an invariable physiology: the sounds people hear may change, and their reactions to those sounds do change, but how people hear remains the same." Schwartz's work is an invitation to open the black box of hearing and attend to auditory variabilities. Indeed, differences in hearing can enable or disable value: one person's auditory capacity may be productive or worthless, or in between, as conditions shift and opportunities arise. The people in our research find the value of their hearing to be modular, and they modulate its value strategically. And because disability, and deafness more specifically, have increasingly become legible as categories in countries associated with the global South because of development intervention, disability has become a key site for parsing questions around the work that categories do and the values that they hold.

## DISABILITY AND DEAFNESS ON A GLOBAL SCALE

At the level of global institutions, disability tends to be defined universally. From 1982 to 1993, the United Nations proclaimed a worldwide Decade of Disabled Persons, which was followed from 1993 to 2002 by a regional Decade of Disabled Persons in the Asia and Pacific Area. Subsequently, the United Nations Convention on the Rights of Persons with Disabilities (UNCRPD) was drafted and ratified by nations around the world from 2006 onward. India and Thailand ratified it in 2007.

However, local conditions afford very different possibilities and limitations for people with disabilities, in fact calling into question the universality of the term "disability." Within Thailand, there are numerous sign languages in addition to the codified National Thai Sign Language, producing communicative zones that do not necessarily align with national borders. Moreover, universalist disability legislation did not arrive until recently and has been uneven—the Persons with Disabilities' Quality of Life Promotion Act replaced the (troublingly named) Rehabilitation of Disabled Persons Act only in 2007. Complicating matters further, a military coup in 2014 suspended the constitution, which had contained antidiscrimination mandates, as well as provisions ensuring access to social services for people with disabilities. Similarly, in India, while a landmark law called the Persons with Disabilities (Equal Opportunities, Protection of Rights, and Full Participation) Act was passed in 1995, Indian Sign Language (ISL) and access to communication were not mentioned, and the law was poorly enforced. The act also applied only to the public sector, which is problematic because employment, education, and services are increasingly provided by the private sector. In the aftermath of signing and ratifying the UNCRPD, a revised law called the Rights of Persons with Disabilities Act 2016 was passed. This Act does mention signed languages; although ISL by name, Indian Sign Language has not been recognized as a language. And while India's Prime Minister Narendra Modi has embraced disability issues, there has been little or no improvement in accessibility, education, or employment. Thus, in both Thailand and India, disabled people are required to navigate a decidedly unstable political-legislative framework.

Development organizations have nevertheless focused on the category of "disability," with increased funding available from both state parties and civil society (what Friedner and Osborne [2015] refer to as "disability development"). This funding seems to accrue in urban areas, where internationalist ways of being disabled are prevalent, and is associated with what the American historian Katherine Ott (2014) calls "disability things" such as ramps, curb cuts, accessible bathrooms, and tactile paths. In the long shadow of such development initiatives, anthropologists and other social scientists have attended to how disability experiences vary depending on political, economic, and social contexts, questioning whether there is a universal disability or deaf experience, in much the same way that scholars have questioned the universality of sense experience (see, e.g., Classon 1993; Geurts 2003). For example, Friedner and Osborne (2015) have written about the failure of universalist access standards in urban India, while Cassandra Hartblay (2017) argues that the proliferation of inaccessible wheelchair ramps (that are too steep and obstructed by steps) in Russia relates to the emergence and rapid enforcement of internationalist standards as builders and contractors hope to "check boxes" in order to satisfy bureaucratic requirements.

The universalist World Federation of the Deaf (WFD) purports to represent deaf people around the world and is part of the U.S.- and Geneva-based International Disability Alliance, which has been involved in drafting and lobbying for the passage of the UNCRPD. Notably, the WFD is a sign-language-focused advocacy organization. Scholars have raised concerns regarding the WFD's espousal of a seemingly normative "right way" to be deaf (Branson

*Friedner and Tausig*

and Miller 2002; Wrigley 1997). These concerns involve a rejection of a medical model of deafness and the recognition of deaf people as members of a linguistic and cultural group. Within the activities of the WFD, sound plays little role. While auditory technologies such as hearing aids and cochlear implants have grown more sophisticated and widespread, the WFD has not focused on them, although it has focused on the emergence of other kinds of technology, such as captioning, video phones, and remote video sign language interpreting services. To be sure, this does not mean that such new hearing technologies are not proliferating. In India, many states provide free cochlear implantation to children living below the poverty line, while nongovernmental organizations around the world are developing low-cost hearing aids. However, in the contexts in which we conduct research, we found that, for a variety of reasons, our interlocutors had chosen not to use these technologies or to use other technologies that they craft or repurpose.

Recent social-science research on deafness has attended to "other ways" of being deaf (Friedner and Kusters 2015; Kisch 2008; Monaghan et al. 2003), and such work has argued for the importance of understanding contexts of deafness—that is, deafness within different constellations of kinship, sociality, and political economy. Of interest to us, and a key way that we see this essay making an intervention, is that there has been little attention paid to the role that sound and listening play in these everyday deaf socialities and to questions of what it means to use the hearing capacities one does have. This absence of focus on audition—in whatever form—is what Tausig is interested in with his attention to salvaged sound, while Friedner is concerned with the ways that sound (and hearing) is spoiled for her interlocutors. We argue that the work done in the fields of deaf studies and disability studies can demonstrate how disabled and deaf people come to have and perform different kinds of value. We now turn to discussions of value and affordances in order to attend subsequently to matters of spoilage and salvation.

POSITIONS, VALUES, AFFORDANCES

While audition does not seem to be particularly valued in the worlds within which Friedner works—it is spoiled—audition is quite important in Tausig's research context, where it is salvaged. While our interlocutors in this particular essay are commensurable in that they all have experiences of deafness to some extent, different kinds of value are created from sound and unsound, audition and inaudition. To think value, we turn to the anthropologist David Graeber (2001: 45), who argues for an understanding of value, in broad terms, as fundamental to world making. Graeber suggests that value is formed through creative action, which congeals as relationships, things, and localities. In figuring value as creative action, we consider how our interlocutors produce, use, and work through sound (or do not) and how this enables the creation of distinct economic and social worlds for them. Just as value as a category always seems to exist in (uneasy) relation to commodification (Graeber 2001, 2013), so, too, do we see the social and the economic in relation to each other in our respective cases.[3]

In addition to Graeber, we invoke affordance theory, which is similarly concerned with creative action and the activation of value. Affordance theory investigates how objects tend to possess or foreclose different utilities (Gibson [1979] 2014; Winner 1986). For instance, an analyst of affordances might note that a wine bottle holds wine well today and can easily be reused tomorrow as a vase but will probably never be valuable as a toothbrush; the wine bottle, through its form and qualities, does not readily afford this last possibility. Moving beyond objects, scholars of affordance theory have asked how a story, conversation, relationship, or academic theory might be thought of as having affordances (Hutchby 2001; Joerges 1999; Nagy and Neff 2015). As Webb Keane (2016: 30) writes, "We might start with the simple observation that not just physical objects but *anything* that people can experience, such as emotions, cognitive biases, bodily movements, ways of eating, linguistic forms, traditional teachings, or conventional practices possesses an indefinite number of combinations of properties." Indeed, Keane (2016: 27) introduces the concept of ethical affordances as "any aspects of people's experiences of themselves, of other people, or of their surround, that they may draw on as they make ethical evaluations and decisions, whether consciously or not."

Affordance theory is a useful tool for an ethnographic examination of the values afforded by hearing and deafness. Hearing is not simply present or absent, partial or complete, in a human sensorium. Rather, hearing is a capacity that acquires or sheds value in its networked engagement with the world. We focus on how the strategic use and nonuse of capacities, such as hearing, may be modulated to draw out value.

SPOILED SOUND

In this section, I (Friedner) consider how mostly middle-class young deaf adults who use sign language in Bangalore conceptualize sound as potentially "spoiled" or, at the very least, unnecessary for sociality. By contrast,

*Friedner and Tausig*

focusing on sign language and deaf social practices "saved" them. These are emic categories that my interlocutors used in a range of settings to refer to the importance of being around other deaf people and of deaf sociality. My attention is to what Karis Petty (2016: 174) terms "sensorial emplacement" to mark the affordances both of the environment and of the individual. In this case, I look at the complex relationships between individuals and the environment to examine how "spoiling" and "saving"—in relation to sound—are modulated. I draw on ethnographic research conducted in Bangalore and other Indian cities with deaf young adults, in which I analyzed their attempts to create inhabitable presents and futures through value-making practices.

I first turn to the scene of an effort to distribute hearing aids among low-income families to provide background on the subject of hearing aids in India (although more recently there has been an increase in cochlear implantation programs targeting low-income families as well). In August 2007, I observed an information session for hearing aids conducted by community-based rehabilitation workers at the nongovernmental organization (NGO) Disabled Peoples' Association (DPA) based in Bangalore.[4] About fifteen people were in attendance, mostly mothers, along with two fathers and some children of varying ages. Most attendees were from impoverished housing colonies and slums located in close proximity to DPA, and they had been found by rehabilitation workers going door to door in their locality looking for disabled children. We sat around a table under ceiling fans in one of the NGO's classrooms listening to the rehabilitation workers extol the virtues of hearing aids, particularly the Siemens body hearing aid that DPA helps parents obtain. A body hearing aid is a rectangular, one-size-fits-all device that amplifies sound; it is hung around the neck or placed in a pocket and has earbuds attached to wires that are plainly visible. In the beginning of the meeting, one of the workers divided attendees into two groups and gave them pink butcher paper and markers. They were told to engage in a discussion about how "the machines" can help their children. (One group wrote in Kannada and the other in English.) The groups both wrote that the machine can help with street safety, with hearing parents at home, with socializing with "normal kids who make fun of them," and with success in school. They also wrote that with the machine, children can learn to talk and to study at a higher level.

This participatory exercise was followed by a demonstration in which the pocket hearing aid was passed around for all to see. The two workers then held up a glossy brochure of new Siemens products, featuring small behind-the-ear and digital hearing aid models, although these models were prohibitively expensive for those present. The Disabled Peoples' Association had arranged to purchase the pocket models, and parents were asked to contribute something to the cost; after the orientation, the two workers met with mothers to determine their financial situation and how much they could afford to pay toward the pocket hearing aids. In one meeting, a mother had two children for whom she wanted to get hearing aids. The family was earning below-poverty-line wages, and the mother wanted to pay 1,000 rupees (around $18) for both hearing aids, an amount that the rehabilitation worker protested was too low. I do not know whether these children ultimately got their pocket hearing aids, but I do know that the models being offered were antiquated and did not work well. In fact, it was odd to me that discussions were being held about the monetary value of these hearing aids when the children had not even tried them. While this was a very specific meeting targeting poor parents, the body hearing aid is part of most of my interlocutors' histories and figures deeply in their memories of childhood education, as I discuss later.

In July 2009, I had a meeting with a senior audiologist named Deepa Sentilraj at the All India Institute for Speech and Hearing (AIISH) in Mysore, Karnataka, at which I told her about my research. In return, she told me that in Bangalore a generation of young adults had not learned to use their residual hearing and were not prescribed proper hearing aids. The AIISH was trying to get the government to cover more up-to-date, behind-the-ear or other hearing aids that were *not* body aids. Sentilraj described the difference as riding a bullock cart compared with riding in a car: both reach the same end point, but the bullock cart driver is much more tired. She said that many states provide cochlear implantation for deaf children and that politicians and philanthropists donate cochlear implants to poor children. (A senior government minister had told me earlier that the central government was involved in trying to manufacture low-cost cochlear implants.) However, there has been very little effort at the national level to develop more technologically advanced hearing aids; presumably, such aids are less impressive than cochlear implants and do less to demonstrate the state's science capabilities.

In our meeting, I told Sentilraj that most of my young adult interlocutors had negative feelings toward hearing aids, and toward body hearing aids in particular. I asked her why she thought this was so, and she replied that "the pocket aids are given out like candy or sweets." Indeed, many of my interlocutors had stories about going to a camp or program at which body aids were given out by politicians or prominent people. They told me that they in-

*Friedner and Tausig*

variably left the camps disappointed.[5] Some wished that more sophisticated hearing aids had been given out and others wished for no hearing aids at all. I was struck by my interlocutors' consistent aversion toward body hearing aids and hearing aids in general. Most told me stories about being forced to wear body aids as small children, which made them uncomfortable and that often did not work well. In addition, these hearing aids marked them as visibly different and provided "normal kids" with a way to make fun of them (in contrast to what the parents involved in the information session mentioned earlier believed). Their earlier experiences with body hearing aids "spoiled" the possibility of using and working through sound, and my interlocutors frequently invoked these earlier experiences as evidence for why other kinds of hearing aids were not of interest or of use to them.

However, the decision not to wear hearing aids was not a resolute one, and my interlocutors modulated in their choices. In addition, visible hearing aids provided young adults (and other deaf people) with a way to meet one another. For example, two young women named Asha and Nandini met while riding the bus. Asha was standing; Nandini saw her behind-the-ear hearing aids and asked whether she was deaf.[6] Asha replied yes and asked Nandini the same question. The two became close friends. Note that Nandini asked Asha whether she was deaf, not hearing-impaired, and their conversation took place in a mixture of ISL and spoken English. For Nandini and Asha, sign language and interacting with other deaf people was extremely important to their development as people in the world. My deaf interlocutors told me that they preferred to spend time with other deaf people, and those who attended schools, vocational training programs, and workplaces where other deaf people were to be found told me that they were unhappy on holidays or vacation periods when they were forced to spend time with their families, whom they could not hear or understand. Deaf people who worked alongside other deaf people often arrived at work early or stayed late to socialize and spend time with other deaf people. What is important to stress, too, is that my interlocutors placed little economic value on sound (although hearing employers and supervisors sometimes did, in occasionally contradictory ways, as I discuss later). Many of my interlocutors told me that they wanted to become teachers or other kinds of educators working with deaf children and adults. Becoming a sign language teacher was a particularly valued profession. Deaf people often told me that they wanted to set up their own NGOs or computer training centers for other deaf people. My interlocutors were thus trying to carve out career paths in which sound afforded little or no economic value, in which audition was effectively spoiled as a value.

However, the Rehabilitation Council of India, the central body overseeing government schools, does not allow deaf teachers in government schools, and while NGOs often employed deaf people, there was concern that they were being paid less than hearing employees. As a result of increased privatization, deaf people increasingly find jobs in business process outsourcing (BPO) and in the growing hospitality sector in India. Many of my interlocutors worked as data entry operators and at coffee cafés and fast food chains. A hearing supervisor at a BPO office told me that the deaf workers he supervised were better than hearing workers because they could not simultaneously gossip and type. In these workplaces, deaf employees' coworkers told me that they found their deaf colleagues inspiring and that their presence translated into positive sentiments directed at their employers. Café Coffee Day, a pan-Indian coffee chain, has developed the category of "silent brewmaster" as it hires many deaf people to be baristas. According to Café Coffee Day executives, deaf people's other senses are better than nondeaf people's senses (see Friedner 2013). Thus, the absence of hearing affords value for corporations that are able to extract economic advantage from deaf peoples' labor. In other words, corporations benefit from the modulatory process whereby sound becomes spoiled for deaf people. Deaf peoples' distinct embodiment combined with the legal-juridical framework surrounding deafness results in specific regimes of valuation that enable both the formation of deaf sociality and the extraction of economic value. We modulate now to Tausig's voice.

### SALVAGED SOUND

The most talented audio technician in Thailand may be a partially deaf, middle-aged father of three named Sam.[7] Born and raised in Bangkok, Sam began tinkering with electronics when he was five, repairing radios in his parents' home. As he and his wife, Hang, tell the story, after Sam lost most of his hearing due to a childhood illness, his impoverished mother and father decided that a technical trade would be better than a high school education for a deaf child and, when he was fourteen, moved him from public school to a vocational institute. By sixteen, in 1982, he was working as an apprentice in one of the best-known car audio shops in Thailand. He earned a national reputation for his skill, and by the end of the 1980s the vehicles he worked on were routinely profiled in Thai-language professional car audio magazines such as *Car Stereo* and *Sound Off*. Sam opened his own garage, SFX Car Audio, in 2007. His renown was such that a high-ranking Democrat politician, Sukhumbhand Paribatra—who would be elected governor of Bangkok two

*Friedner and Tausig*

years later—cut the ribbon at the garage's opening ceremony. Sam's reputation and business success attest to his exceptional skill as an installer. The SFX garage is today a destination for both Thai and foreign car audiophiles seeking state-of-the-art sound systems for their vehicles.

Sam supervises three or four employees in the shop each day, with the help of Hang as an office manager. Relative to other garages, SFX is medium-size, with two parking spaces inside and room for a third and fourth car on the patch of sidewalk outside. Sam's employees are trained to be exacting with every wire and connection, and their reputation precedes them. Even in a difficult economy after the economic crisis of 2008, Sam's business has remained successful.

Sam can hear loud sounds and voices in close proximity or in isolation; he is partially hearing or, in Thai, *hu dteung*. (He describes his capacity in just such spatial terms—perhaps we should call him remotely deaf/proximately hearing). This enables him to test vehicular audio with his ears under controlled conditions, but it also requires him to supplement his hearing with other forms of testing. For example, by examining the strength of soldering connections or running his fingers over the interior panels of a vehicle while music plays, he can tell whether a speaker system has been installed to his own high standards. As he works, he modulates in and out of auditory engagement, shutting sounds out and bringing them back to his attention once again, balancing sonic details as inaudible reference points while he proceeds haptically and visually.

In May 2011, I observed Sam as he soldered wires to an amplifier in the trunk of a BMW 323i. He sat on the lip of the trunk, legs hanging and torso turned inward. The door of the trunk shielded him from the street. His hands moved rapidly as he rearranged wires, a small kit of tools at his side. Sounds entered the garage at multiple frequencies from the busy four-lane highway outside: wheels on asphalt in the middle range, a constant line of sharply squeaking brakes in a higher register, and the occasional scream of a motorcycle revving up. Within two minutes, Sam had finished. He lay the joined wires casually on the soft felt interior and gave a quick look of satisfaction as he hopped down and moved on to his next task.

Hang often narrates such situations on Sam's behalf, perhaps because he is shy, although when they do speak together, it is clear that they agree on narratives. After Sam finished soldering, she explained,

> It's as if he can't hear noise. He can do complicated (*sápsɔ́ɔn*) modifications. Do you know the word *sápsɔ́ɔn*? It means complex. Car sirens.

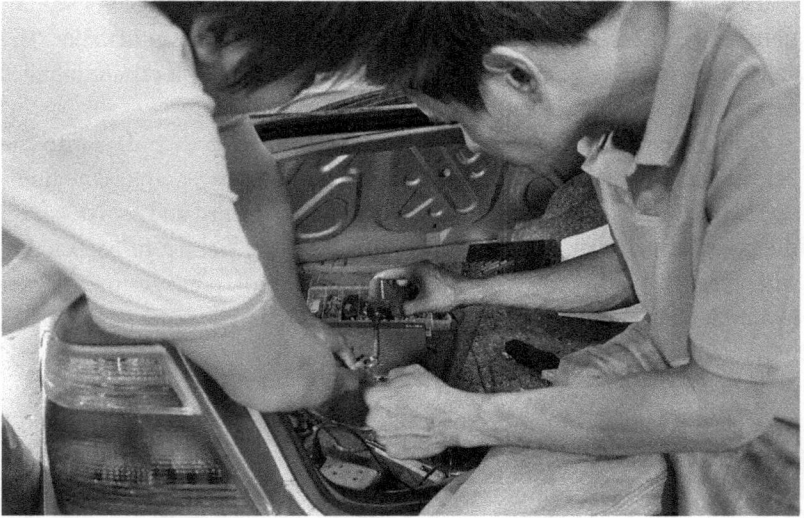

FIG 6.1. Sam and an employee work on audio connections in the trunk of a BMW in Sam's shop. Photograph by Benjamin Tausig.

*Bee-nee-bee-nee-bee. Boom!* Sam doesn't hear them. Therefore, Sam can work 100 percent. Focus 100 percent. If you can hear *bee-nee-bee-nee*, you can't work. You get distracted. You lose your concentration. Your focus is lost. But Sam's focus is OK. We call it a heavenly gift (*phɔɔn sàwǎn*).[8]

What kind of a "gift" does Sam possess? What is afforded by the possession of such a gift in the context of Sam's garage, in Bangkok, and in this particular node of cosmopolitan global exchange? Conceptualizing Sam's proximate hearing/remote deafness as neither lack nor physically determined capacity but, rather, as a modular asset in a particular field of exchange leads toward a richer conception of value. Sam exploits the affordances of his senses. He is not the idealized unhearing person who cunningly judges sound with his eyes, swapping one sense for another, but an expert with a range of sensitivities and capacities (as well as limits), who sees, hears, and feels not just sound but sound *systems*. We might in fact compare Sam's auditory value to the architecture of a car audio system, which is knotty and multiform, having components in nearly every region of the vehicle, some visible and accessible and others buried within doors or behind interior panels. Some parts of a car audio system emit or receive sound directly, while others control, modify, or transmit it remotely or indirectly. Architectures of auditory value

*Friedner and Tausig*

might be imagined in similar terms. Although hearing's value may appear to emerge sui generis in the moment of audition, such value depends on intricate systems, whose components are not nearly all related to the sonic. Sam's engagement with audition is modular, tactical, and always integrated with nonsound.

The customers who take their cars to Sam at SFX expect to hear music in effective isolation. They pay large sums of money so their car-based listening will be an impenetrable auditory experience, a common desire for elites in many parts of the world (Bijsterveld et al. 2013). Although perfect aural solitude is impossible in practice, high-end car audio enthusiasts in Thailand idealize sound systems that seem to approach total personal immersion. To whatever extent possible, they wish to seal off the privacy of their own sense worlds. One might expect Sam to encounter skepticism or even outright bias from potential customers. However, in Thailand's prevailing climate of high-end listening, Sam's proximate hearing in fact gives him an advantage as a technician—it is valued. His customers agree with Hang that because Sam can hear in the near field but not from far away, he is especially able to focus on the sound of a car audio system while avoiding distraction from urban noise in a way that a fully hearing person could not. The kind of hearing that his "heavenly gift" affords links profitably with the preferences of the most discerning audiophiles for listening in isolation. It is no surprise, then, that the bulk of SFX's clientele are wealthy elites (most with links to the global North), commissioning fantastically expensive stereo modifications to listen lavishly, while shushing the rusty hack and grind of a far-from-post-industrial city. Sam as an artisan seems to embody the sensory stratification that his customers desire, and he is in turn proud to have refigured, or salvaged, what might otherwise be construed as a disability into a tool of art, a heavenly gift.

The systems Sam installs are meant not for parties, concerts, or other forms of public listening but for individuals. This emphasis on interiority contrasts with a far more prevalent and financially low-end culture of car audio in Thailand in which young men flaunt flashy stereo systems installed on the cheap. Industrial areas not far from Bangkok, such as Chonburi and Rayong, host trade shows that feature audacious car and truck sound systems. There is a strong participatory character within Thailand's low-end car audio communities. Sam derides the shops that install these systems, claiming that their work is shoddy. In particular, he notes that the joints between audio cables are often crimped rather than soldered together, or they are soldered poorly, undermining the clarity of the signals and introducing noise. The shops themselves are dirty and disorganized, he says, with clutter everywhere. At SFX, as Sam proudly shows off to visitors, everything is in its proper place. All of his hardware is labeled and neatly arranged in cases in the rear of the garage.

Among Sam's methods for testing a car audio system is to concentrate the sound emanating from the speakers with a length of insulated blue plastic pipe. He holds this pipe up to his ear as music plays inside the car, allowing him to assess it for any potential flaws or imbalances. His testing is the final word, and if he detects a problem, his employees will fix it, even if that means a half day's work of removing panels to reach a faulty connection. The ideal method of testing is one in which a sound signal can connect to the ear of the tester with minimal interference from exterior sources. The blue pipe stages the most direct possible connection between source and ear, the purest available isolation. Sam needs the pipe to concentrate the sound tightly enough to hear loudly, but it serves the simultaneous purpose of purging the signal of unwanted noise. The stratification that occurs here is not a given condition but one that must be produced and sustained through a combination of practices, devices, and physical capacities.

Sam has earned, over time, a very good living. His audio installation jobs may cost as much as two million baht ($70,000). Moreover, he has been invited to factories in the United States and China as a consultant for companies designing new car audio products. His customers, he estimates, are 20 percent foreign, and the vehicles that come through the garage are nearly all late-model BMWs, Toyotas, and Mercedes-Benzes. By succeeding in a business with a wealthy and cultured clientele, Sam has moved from poverty in his youth to an upper-middle-class lifestyle as an adult. Sam and his family take weekend excursions to high-end shopping malls such as K-Village and Gaysorn Plaza, where one can drink Italian wine and browse at the Burberry store. These malls have parking garages and direct entrances from the above-ground mass transit system, allowing customers to travel from home to shopping without a moment of exposure to the infamous noise and air pollution of the city. Thanks in part to Sam's heavenly gift, SFX's business has boomed, and Sam has achieved upward mobility.

The ideological language of the upper class, imagined as a unit of the global North, is not always kind to those outside of it. One afternoon I chatted with a young Thai man named Jack who went to college in California and who was having minor sound work done in his car. Jack dismissed Thailand's poor as brainwashed. He described a divide between educated and uneducated, poor and wealthy in the country. For the professional and educated

*Friedner and Tausig*

class, condemned to torturous commutes, he argued, expensive car audio is a daily necessity. Sam and Hang are much more tactful than Jack. However, their clients have little sympathy for the poor. The hermetically sealed sonic interiors they desire are meant to appease a sensory discomfort with poverty, environmental degradation, dirty industry, and municipal incompetence.

Sensory capacity has little meaning prior to the world in which it is used. Capacities — or, in Hang's language, "gifts" — are always unique rather than universal. Moreover, they are subject to processes of modulation in which their value emerges and disappears, is activated and extinguished. Sam's deafness is constituted by those around him as a lack, an abnormality, at the same time that it is made into a "special gift." His partial or proximal deafness is a justification for abandoning high school as well as a technique for class mobility. These consequences became possible in Bangkok in a particular period of history, forged in a special crucible of prejudices, opportunities, and mediating technologies. Hearing must be freshly theorized as local conditions demand. Figuring hearing's value as subject to modulation may contribute to these theorizations.

### CONCLUSION: A MODULATED ENDPOINT

Sensory capacities are not biologically determined before a person steps into a network of cultural projects and local distinctions. Rather, sensory capacities *emerge within* social, political, and economic contingencies. At times, this means that a seeming disability might function in practice as an asset and that a seeming ability might be worthless, even detrimental, to a person or to a community's experience. Indeed, even these functions and their values modulate as conditions demand. In each empirical case, hearing and deafness must be situated and understood as processual before we can understand their utility.

To discuss such situatedness, we focused on two ethnographic examples, emphasizing value as a way to understand how ability and disability function in subtle ways. Hearing may be spoiled where it is perceived to have little or even negative value, while ostensibly damaged or "partial" auditory capacities may be salvaged when understood as productive. Recourse to affordance theory, particularly recent treatments that engage social abstractions as material agents and that examine the ways that affordances can also be ethical, open a path for thinking about ability and disability not as innate capacities but as strategic decisions in fields of value — not economic value alone but, rather, the richly plural value systems that course through social processes of all kinds, that, as Graeber has it, create worlds. Our interlocutors both emphasize and strongly resist the binary of hearing and deaf in their strategic decision making. Furthermore, we argue that it is important to move away from an overly rigid, deterministic hearing-deaf binary that seems to drive the work of international disability and deaf organizations that mobilize universalist definitions of disability as both individual condition and social problem. Note that these agencies, though linked rhetorically to the global North, are not fixed on a map; nor are the people whom they address. Movement between the global North and the global South might also come into view in thinking about binaries as subject to modulation. To exploit the value-laden affordances of sensory capacities is also, as our research suggests, to navigate deftly among channels of status and opportunity, the very channels that the global North-global South binary would hold as distinct.

### Notes

1. "The spoiled and the salvaged" is a binary, similar to those explored in Lévi-Strauss 1983, among other structuralist works of anthropological theory. Like Lévi-Strauss, we believe that "empirical categories—such as the categories of the raw and the cooked, the fresh and the decayed, the moistened and the burnt, etc., . . . can only be accurately defined by ethnographic observation" (Lévi-Strauss 1983: 1). Unlike Lévi-Strauss, however, we seek an alternative to universal laws and to binary systems that find permanent resolutions.

2. We think athwart biology much as Stefan Helmreich (2011: 136) suggests thinking athwart theory, "neither as set above the empirical nor as simply deriving from it, but as crossing the empirical transversely . . . an abstraction as well as a thing in the world."

3. In this vein, disability studies scholars (e.g., Barnes and Mercer 2010) have argued that disability as a category emerged in relation to political-economic changes that accompanied urban industrialization. As modern factories with assembly lines emerged, so did disability as a category. Disability can thus be considered an economic category.

4. The names of all NGOs and individuals have been changed in Friedner's section.

5. Such camps are not only held in countries in the global South. For example, the Starkey Foundation held a similar camp, albeit with more technologically advanced hearing aids, alongside the 2016 Super Bowl. This camp caused a furor in American deaf worlds as attendees reacted negatively to the focus on deafness as needing cure. They also resented having celebrities (including the rap star 50 Cent) fit their children with hearing aids as journalists took photos.

6. Asha was from an upper-middle-class family, and her father and sister were both "hard-of-hearing" (they did not identify as deaf). All three of them wore behind-the-ear hearing aids, and Asha attended an oral school where she learned to speak and lip-read. However, when she entered college, she felt increasingly isolated from

*Friedner and Tausig*

hearing peers and started learning to sign and sought out the company of other signers. While she strategically talks and lip-reads with her family and while navigating through the city, she prefers to sign. Many of Friedner's interlocutors used speaking and lip-reading skills when necessary, if they had them, and drew on a wide range of communicative skills across speech, sign, and gesture.

7. Tausig met Sam in 2011 as part of a larger project to examine how car stereo systems were installed for use at political rallies. Sam (along with his family, employees, and customers) is likely to discuss his deafness as not an absolute condition. No medical or audiological measurement was ever offered; nor did such a measurement seem of interest. The term "partially deaf" does not correspond to a precise degree of hearing loss or capacity. Rather, it is a summary term derived from Sam's own characterization of his own hearing and nonhearing.

8. It is difficult to translate *phɔɔn sàwǎn*. Literally it means "heavenly blessing," but idiomatically it means something more like "special talent," without strong religious connotations.

## References

Barnes, Colin, and Geof Mercer, eds. 2010. *Exploring Disability*. Cambridge: Polity.

Bijsterveld, Karin, Eefje Cleophas, Stefan Krebs, and Gijs Mom. 2013. *Sound and Safe: A History of Listening behind the Wheel*. New York: Oxford University Press.

Branson, Jan, and Don Miller. 2002. *Damned for Their Difference: The Cultural Construction of Deaf People as Disabled*. Washington, DC: Gallaudet University Press.

Classon, Constance. 1993. *Worlds of Sense: Exploring the Senses in History and across Cultures*. New York: Routledge.

Friedner, Michele. 2013. "Producing 'Silent Brewmasters': Deaf Workers and Added Value in India's Coffee Cafés." *Anthropology of Work Review* 34, no. 1: 39–50.

Friedner, Michele, and Annelise Kusters, eds. 2015. *It's a Small World: International Deaf Spaces and Encounters*. Washington, DC: Gallaudet University Press.

Friedner, Michele, and Jamie Osborne. 2015. "New Disability Mobilities and Accessibilities in Urban India." *City and Society* 27, no. 1: 9–29.

Geurts, Katherine. 2003. *Culture and the Senses: Bodily Ways of Knowing in an African Community*. Berkeley: University of California Press.

Gibson, James. [1979] 2014. *The Ecological Approach to Visual Perception: Classic Edition*. London: Psychology Press.

Graeber, David. 2001. *Toward an Anthropological Theory of Value: The False Coin of Our Own Dreams*. New York: Palgrave Macmillan.

Graeber, David. 2013. "It Is Value That Brings Universes into Being." *Hau: Journal of Ethnographic Theory* 3, no. 2. Accessed April 1, 2016. http://www.haujournal.org/index.php/hau/article/view/hau3.2.012.

Hartblay, Cassandra. 2017. "Good Ramps, Bad Ramps: Centralized Design Standards and Disability Access in Urban Russian Infrastructure." *American Ethnologist* 44, no. 1: 9–22.

Helmreich, Stefan. 2011. "Nature/Culture/Seawater." *American Anthropologist* 113, no. 1: 132–44.

Hutchby, Ian. 2001. "Technologies, Texts and Affordances." *Sociology* 35, no. 2: 441–56.

Joerges, Bernward. 1999. "Do Politics Have Artefacts?" *Social Studies of Science* 29, no. 3: 411–31.

Keane, Webb. 2016. *Ethical Life: Its Natural and Social Histories.* Princeton, NJ: Princeton University Press.

Kisch, Shifra. 2008. "'Deaf Discourse': The Social Construction of Deafness in a Bedouin Community." *Medical Anthropology* 27, no. 3: 283–313.

Lévi-Strauss, Claude. 1983. *The Raw and the Cooked.* Chicago: University of Chicago Press.

Monaghan, Leila, Constance Schmaling, Karen Nakamura, and Graham Turner, eds. 2003. *Many Ways to Be Deaf: International Variation in Deaf Communities.* Washington, DC: Gallaudet University Press.

Nagy, Peter, and Gina Neff. 2015. "Imagined Affordance: Reconstructing a Keyword for Communication Theory." *Social Media and Society* 1, no. 2: 1–9. doi: 10.1177/2056305115603385.

Ott, Katherine. 2014. "Disability Things." In *Disability Histories*, ed. Susan Burch and Michael Rembis, 119–35. Urbana: University of Illinois Press.

Petty, Karis. 2016. "Walking through the Woodlands: Learning to Listen with Companions Who Have Impaired Vision." In *The Auditory Culture Reader*, 2d ed., ed. Michael Bull and Les Back, 173–85. London: Bloomsbury.

Schwartz, Hillel. 2011. *Making Noise: From Babel to the Big Bang and Beyond.* Cambridge, MA: MIT Press.

Winner, Langdon. 1986. "Do Artifacts Have Politics?" In *The Whale and the Reactor: A Search for Limits in an Age of High Technology*, ed. Langdon Winner, 19–39. Chicago: University of Chicago Press.

Wrigley, Owen. 1997. *The Politics of Deafness.* Washington DC: Gallaudet University Press.

# 7

# Remapping the Voice through

# Transgender-*Hījṛā* Performance

*Jeff Roy*

. . . . .

The voice is understood as a multifaceted set of practices within which orality, aurality, and subjectivity are conceptually and experientially interlinked. In contemporary India, as in many parts of the world, however, the privileging of certain virtuosic vocal and performance techniques has been, and continues to be, interlinked with the interests of nation and capital (Weidman 2006) and reinforced through the subjugation of certain cultural practices and identities that work against these interests (Soneji 2011). The processes of validating and authorizing vocality are invariably linked to the marginalization of communities from which these voices originate. In cinema, television, or even staged performances, for instance, the image and sound of the singing and dancing *hījṛā* (otherwise known as "third gender") often work to index the communities' vulgarity, deviance, or social impotence, rather than their legitimacy as artists in their own right.

In this chapter, I reflect on how my sonic engagements with the religiously and linguistically multifaceted transgender and hījṛā (or what can be termed trans-*hījṛā*[1]) communities in India produce different understandings of voice. I put forth the claim that in trans-hījṛā—and, indeed, other—contexts, the voice and its correlative identities should be understood outside the determinative framework of virtuosity and within the framework of *izzat* (roughly translated as "respect"), since it pertains to voices and identity expressions that elude the logics of gender in which national and transnational identities are exchanged. Put simply, trans-hījṛā communities sing—or otherwise "sound out" through uniquely stylized nonvirtuosic vocalic practices—as a means of generating respect among their members and to transcend normative sonic spaces that engender normative behavior and identities. Situated at sound studies' recent turn toward the global South, following the anthropologist Gayatri Reddy's (2005) formative ethnography on the izzat of hījṛās in Hyderabad and joining Aniruddha Dutta's and Raina Roy's call to decolonize hījṛā from the discourses and cultural practices that frame regional and otherwise "local" gender nonconforming identities (see Dutta 2016; Dutta and Roy 2014), this essay seeks to remap our understanding of identities that for so long have contested or ignored conventions of aural approval.

The chapter also follows a collaborative article produced by the performance studies scholar Pavithra Prasad and myself on the need to embrace a performance-based scholarly practice within the framework of "sincerity" rather than authentic and virtuosic technique (vis-à-vis John Jackson's [2005] work on "racial sincerity"). As Prasad and I suggest, music scholarship's preoccupation with virtuosity as an interpretive regime reproduces the very same constraints on which classical musical training has historically relied to protect its dominant status in colonial and postcolonial India (Prasad and Roy 2017).[2] Taking a cue from the performance studies scholars Ramón Rivera-Servera (2009) and Alejandro L. Madrid (2009), in this follow-up essay I seek to address how vocal performance in particular opens up its subjects, especially in cases where such performances lead to experiences and understandings of gender that lie outside forms, techniques, and compositions that classical musicians have been trained to master. By releasing the singing voice from the ironclad constraints of virtuosity, queering conventions of artistic production and knowing (Halberstam 2011), and embracing the imperfect and sometimes deviant vocalizations of the body, I seek to mitigate the colonial hangovers in scholarly practice and achieve a more inclusive and multivalent hermeneutic for the study of gender and its many applications.

RECOGNIZING VIRTUOSITY AT ARM'S LENGTH

The coconstitutive relationship between gender and voice is explored in Amanda Weidman's influential *Singing the Classical, Voicing the Modern* (2006), which answers the call for an alternative approach to the study of the voice as framed by the classical. Her work exposes the South Indian classical tradition as a modern construct and as part of a nationalist agenda that not only reoriented the way performers produced and listened to music, but also authorized new codes of gendered conduct. Around the time of inde-

*Jeff Roy*

pendence, the Brahmin female voice became the emissary of an "authentic" Indian identity as new values of femininity that privileged middle-class respectability, chastity, and domesticity were inscribed onto the bodies of Indian women. This strategically situated India in opposition to the West and created a classical music landscape defined by originality (rather than reproduction), humanity (rather than mechanization), and tradition (rather than modernity) (Weidman 2006: 5). The implication here is that listening to and sounding out the virtuosic voice attuned—and, indeed, still attunes—people to gendered behaviors infused with and subdued by class norms, sanitized worship, the restraint of sexuality, and patriotism.

In some ethnomusicological literatures of the global North, the voice has been defined as a polysemic social practice that serves as an embodied performance of the performer's sense of self and of community, and it is treated as an experiential bridge linking culturally specific virtuosic performances with an emergence of self-understanding. In the frequently cited work by Jane Sugarman, for instance, the voice—or the practice of singing, more specifically—is understood both to contribute to "the ongoing consolidation of practices" that produce the communities from which they originate, as well as to define the dialectic space between voice (or singing) and subjectivity, where the self is manifested experientially (Sugarman 1997: 3). Following this line of logic, a virtuosic vocal performance becomes integral not only to the representation of the musical body in relation to other bodies in communal contexts, but also to the conception and expression of gender identities as they are negotiated into being through these performances.

In work on transgender singers, the relationship between vocal virtuosity and gender is slightly revised. Elias Krell's study of transgender singers in Brazil, for instance, contends that the voice "literally and figuratively speaks affective trajectories that offer critical insight into the ways in which transgender subjects experience and negotiate identities and bodies" (2013: 489). As the voice shifts from higher to lower registers, as it did, for example, for the trans-male singer Lucas Silveira, it "enacts what has been called transgender vocality, or trans-vocality, in that [it] works to unhinge certain affects from their assumed correlative genders, and prescribed genders from presumed bodily morphologies" (Krell 2013: 495). While the voice unhinges affects from their assumed correlative genders, the demonstration of virtuosity nevertheless remains integral to the authentication of a gendered identity, even when—or, perhaps, even more so when—that identity does not correlate with one's apparent gender. In this case, the illegibility of the professional artist is read as a reflection of their virtuosic devotion to their craft, which, while advantageous, still relies on a legible reading of gender as integral to the making of the self. I would suggest that in hījṛā contexts, even as conceptions of self-understanding are changing alongside negotiations of transgender, hījṛā voice is less attuned to this reflexive construction of gender. Rather, it supposes an evasion of vocalic gendering altogether through a particular and unique irreverence toward virtuosity.

José Esteban Muñoz's (1999) "disidentification" model makes sense of the performances with which queer people of color and other minoritarian subjects "tactically work on, with, and against" dominant cultural regimes. Neither conforming to nor squarely resisting these regimes, queer subjects strategically evade coherent readings through creative forms of identity failure (Halberstam 2011). I interpret trans-hījṛā vocality—or what could be shortened to hīj-vocality—in a similar way, while also keeping in mind nuances particular to their subcultural contexts. Hīj-vocality does not produce a virtuosic reading of gender, nor does it seem to be concerned with authenticity or originality insofar as stable interpretations of these terms are concerned—though this is not to say that trans-hījṛā individuals are not virtuosic or that they do not strive for virtuosity. In my six years of fieldwork with communities throughout India, I have heard many trans-hījṛā vocalists who are expert singers in North and South Indian classical musics. Nevertheless, while virtuosic voices are admired and practiced by individuals in the trans-hījṛā community, the recognition of virtuosity does not serve the same important role in the collective authentication of trans-hījṛā identity. Moreover, while expert singers may be recognized as experts by individuals in the trans-hījṛā community, many of them remain unrecognized as experts in normative contexts or, in some cases, may be targeted explicitly as a result of their virtuosic performance of gender subversion.

One example that comes to mind is the singer Gopi, whom I met at the Koovagam festival in Tamil Nadu in April 2013. Although Gopi does not self-identify as transgender or hījṛā, his vocal performances are deployed on a spectrum of disidentification, achieving in many ways that which hīj-vocality seeks to achieve, albeit with a difference: his use of virtuosic vocal technique. Gopi is a professional playback singer whose vocal performance rests on the exact replication of the female singer through virtuosic manipulations of the falsetto range (a method of vocal production used by male singers to sing notes using the "head voice"). His intonation and tone reflects a virtuosic sensitivity to mimetic accuracy to the original tunes, a sensitivity that can be heard in a short film I published in 2013.[3] In his performance of "Kotta Paakkum" (Betel Nut), for instance, Gopi follows the exact contours of the

*Jeff Roy*

playback singer's voice in her register while another male singing partner performs in a lower register. His playful subversion of gender—and the sexuality made explicit through the wink of an eye—enacts a deliberate misreading of normative gender roles established within the world of Tamil film-music culture. If singing queerly reflects a strategic virtuosity—what Yvon Bonenfant (2010: 78) considers "a virtuosic development of the performance of giving attention"—then Gopi's singing reflects a certain kind of attunement to vocalizing what is conventionally pleasing to the ear in order to queer any sonic recognition of gendered belonging.

In the video, trans-hījṛā audience members praise Gopi by showering him with money. Although he is recognized for his talent, the recognition of his virtuosity extends only as far as the length of the stage. Following his vocal performance, I conducted an interview with him backstage, literally in a closet. In the interview, he did not explicitly "come out" but, rather, spoke in code. "My family doesn't know that I'm like *this*," he told me, "but I am participating in Koovagam for my happiness as well as everyone else's."[4] Gopi's story echoes that of other *koṭhīs*—effeminate or otherwise self-defined "passive" men—who sometimes engage peripherally in the affairs of the trans-hījṛā community and who, most of the time, live double lives. Gopi said, "At home and in public places, I've faced a lot of hurdles, tortures, insults, and shame. But, I don't care about any of that stuff. I stand firm in what I believe in my heart. And now my parents are proud of my singing." Although his virtuosity plays a role in the building of personal, private conceptions of self and selfhood, to protect his physical, emotional, and social stability, it is held at arm's length in the authentication of his offstage gendered identity.

## HĪJ-VOCALITY AND THE EVASION OF VIRTUOSITY

Whereas queer vocality (via Gopi) and trans-vocality (via Silveira) rely on a virtuosity that locates certain prescriptions of gender and subverts them to varying degrees, hīj-vocality seems to perform a particular kind of virtuosic failure. Trans-hījṛā voices are prominently exhibited in the ritualistic acoustic musical practice known as *badhai*, which are customarily performed as performative blessings of fertility and financial prosperity for willing or unwilling patrons at weddings, births, store openings, and other auspicious occasions involving important pecuniary milestones. The small ensembles that perform badhais are often composed of one lead singer, a handful of supporting singers who may dance and clap, a *dholak* (two-faced membranophone) player who provides rhythmic accompaniment, and, at times, a harmonium player who carries the tune. Performed frequently at hījṛā *jalsas* (literally, "meetings") and interregional gatherings, badhais also sustain the socioeconomic vitality of the *gharanas* (households; literally, "of the house") and inter-*gharanedar* community and serve as the basis for the community's primary organizational and pedagogical system, the *guru-chela* (teacher-disciple or, colloquially, mother-daughter) relationship. In this system, learning how to become a hījṛā consists largely of aural osmosis that depends on demonstrations of "service" manifested in the form of izzat (Reddy 2005). Izzat has a large bearing on the ways hījṛā chelas navigate their identities within the gharana, and an ability to perform is what sustains their place in the gharana. In this case, "ability" is marked not by one's attunement to musical virtuosity but by the provision of respect to the badhai troupe.[5]

In my experience interacting and, in some cases, performing with trans-hījṛā ensembles in Uttar Pradesh, Gujarat, and Maharashtra—which I discuss further in several case studies in my dissertation and articles (Roy 2015, 2016, 2017)—vocal performances are not predicated on repetition or replication of virtuosic sounds but instead share several distinct qualities. The singing (and in some cases, speaking) voices are relatively high in register, voluminous, and produce a timbre that can be described as nasal, raspy, and piercing. Songs are generally performed in the *madhya saptak* (middle register above middle C)—the same as in conventionally performed in vocal *sthais* (the primary theme of a *ghazal* performed in the first half of the middle octave) and in middle singing ranges in Indian classical music forms. While the singing voices occupy a middle space somewhere between the conventional male and female ranges, the notes never exceed the limits of either. Singing out of tune is also a common practice, although the guru or lead singer may discourage it. In Mumbai many of the singers I encountered cared little for being "in *sur*" (having correct pitch), while in other locations, such as in areas of Gujarat and Uttar Pradesh, intonation was as uncompromising as in the classical tradition. Differences in singing styles can be heard, for example, in the performances of the same badhai song by three different ensembles from Kalyan, Surat, and Mumbai, respectively.[6]

In contrast to Gopi's performance, the trans-hījṛā singers' timbres and tones reflect an attunement to vocalizing and hearing *beyond* what is classically pleasing to the ear. Their uniformity of voice, although expressed in the classical sense, reflects an intentionality not toward mimetic virtuosity but toward an expression of vocal difference that is meant to cut through the static of normative urban spaces. Tripta Chandola (2011: 4–5) writes about the "sensorial ordering" of certain sounds in urban landscapes and how

*Jeff Roy*

they tend to be identified through a process she calls "listening into others." Through these processes, people create sensorial hierarchies that are used to "segregate, exclude and discriminate." I might also suggest that sound and, more specifically, distinct vocal sounds allow trans-hījṛās to identify one another and to engender respect with others in and outside their immediate community. At least in theory, the more identifiably trans-hījṛā one voices out, the more attention and potential izzat one garners in the situation.

This reminds me of the several encounters I have had with trans-hījṛā groups in the streets and on Mumbai's local trains. In these cases, some trans-hījṛā individuals attracted the attention of prospective patrons by clapping in the signature trans-hījṛā style (both palms flat with fingers splayed) and by projecting their voices using the diaphragm and the narrowing of sound through the nasal cavity. The two techniques combined produce a distinct and unmistakably trans-hījṛā sound that is used strategically to cut through the din of a typical crowd in the attention economy of a performance.

Hīj-vocality encompasses a strategy of garnering respect for their gendered difference from normative society. Unlike Gopi's performance, which rests on the playful vocal inversion of gender roles yet ultimately relies on an ability to conform to a stable notion of what constitutes male or female, hīj-vocality unsettles and disrupts the tonal soundness on which notions of male and female rely and carves a distinct sonic space for itself. The emphasis in performance is not in the virtuosic recitation of male or female vocal patternings but in sounding unapologetically trans-hījṛā. This difference does not necessarily make trans-hījṛā a separate or third gender per se; rather, it makes it an identity that creatively resists legible readings of any kind. As unreadable, hīj-vocality engenders emotional affects, values, and social practices that situate trans-hījṛā identity within a "respectful" strategy of difference from heteronormative culture as well as from Indian queer subcultures, taking one immeasurable step further from Muñoz toward an ideal of complete gender indifference.

## TRANS-HĪJṚĀ DIS-IDENTITIES

The disruption of vocal virtuosity on which normative conceptions of 'male' and 'female' rely reflects a larger strategy of ambiguity that trans-hījṛā communities throughout South Asia employ to prevent intrusion from outside forces. In his work with hījṛā communities in Lahore, Faris Khan (2016) notes how, to thwart the efforts of others to enter and control their gharanas, many communities frame their identities through an essential spiritualism that, through vagueness and ambiguity, is materially unavailable. This is revealed in the hījṛā community's recent adoption of the term *Khwaja Sira*, an Urdu term of respect, which represents a discursive shift that may confuse those on the outside and, in particular, the Pakistani state, which has recently attempted to "conquer" the ambiguity of hījṛā through institutionalization.[7]

Institutional definitions of hījṛā are neither clear nor entirely helpful to the community's efforts toward cultural preservation. The anthropologist Sayan Bhattacharya (2017) has argued that the widely used term "neither male nor female" invisiblizes hījṛās as a definition rooted in the binary to which it never truly conforms. At the same time, while recently introduced parliamentary definitions—following a ruling by the Indian Supreme Court in April 2014 and later by Parliament—seem to offer the conciliation of visibility, they enact another kind of erasure in their definition of the community. As Dutta (2016) has pointed out, the term 'transgender' privileges those who have undergone sexual reassignment surgery, or who have "come out," and invisiblizes pre- or nonoperative transgender individuals, not to mention the hījṛās, *thirunangai*, *siva-sati*, and other "local" identities that do not wish to adhere to the Western ideological frameworks these terms construct. While seeking a definition that is intelligible yet refractory, inclusive yet not universal, particular yet not confined, "local" yet not dependent on the hierarchies of scale established by global discourses and practices, I suggest that we also consider the ways in which identities are defined not only by their adherence to stable constructions—whether vernacular, transnational, or otherwise—but also by their failures to adhere to them in the first place.

Hīj-vocality, if anything, shows us how failure—of heteronormativity, of phenotypical belonging, of cultural and behavioral prescription or adherence, of virtuosic singing—can serve as a useful strategy in the cultural survival of marginalized subjects. In our studies of voice and sound, we must make room for failure so that our subjects are found and recuperated in the field.

## Notes

1. In India, the English term "transgender" has served as a catch-all used for gender nonconforming communities that are otherwise distinct from the hījṛā community. These include *jogtīs* (temple devotees of Goddess Renuka-Yellamma), *śiva-satīs* (devotees of Shiva), and *Aravanis* (devotees of Lord Aravan from Tamil Nadu). However, as I discuss in a recent article (Roy 2016), the word "transgender" has also become incorporated into hījṛā vernacular to signify one's gender identity as self-understanding (distinct from the assigned gender at birth), as it is most often employed in the global

*Jeff Roy*

North. This signification has proliferated alongside the increased availability gender confirmation surgery in Indian hospitals and trans-affirmative policy leading up to and following the April 2014 decision by the Indian Supreme Court to grant legal recognition to the third gender. While some hījṛās currently identify as transgender, others still do not. I have therefore adopted the hyphenated term "trans-hījṛā" to accommodate the shifting uses and meanings of transgender in India.

2. See our discussion of the field's literature in Prasad and Roy 2017.

3. See the web page for the film at http://fulbright.mtvu.com/jroy/2013/06/06/meet -gopi-koovagam-part-2.

4. Gopi, personal communication, April 25, 2013, emphasis added.

5. For a thorough discussion of the gharana and music tutelage system, see Roy 2015.

6. See the web page at http://www.jeff-roy.com/asha-natoru, password: dholak.

7. Pakistan was among the first Muslim states in the world to officially recognize the hījṛā community and employ them in tax collection.

## References

Bhattacharya, Sayan. 2017. "Unhoming the Home as Field: Notes towards Difficult Friendships." Paper presented at the 2nd Queer Preconference: Navigating Normativity from a Non-Normative Perspective, 46th Annual Conference on South Asia, University of Wisconsin, Madison, October 26.

Bonenfant, Yvon. 2010. "Queer Listening to Queer Vocal Timbres." *Performance Research: A Journal of the Performing Arts* 15, no. 3: 74–80.

Chandola, Tripta. 2011. "Listening In to Others: Moralising the Soundscapes in Delhi." *International Development Planning Review* 34, no. 4: 391–408.

Dutta, Aniruddha. 2016. "Vernacularization and the Scalar Hierarchies of LGBT Politics in South Asia." Paper presented at the 1st Queer Preconference: Explorations of Queer Methodologies, Identities, and Local/Global Translations in South Asia, 45th Annual Conference on South Asia, University of Wisconsin, Madison, October 26.

Dutta, Aniruddha, and Raina Roy. 2014. "Decolonizing Transgender." *Transgender Studies Quarterly* 1, no. 3: 320–37.

Halberstam, Judith [Jack]. 2011. *The Queer Art of Failure.* Durham, NC: Duke University Press.

Jackson, John L. 2005. *Real Black: Adventures in Racial Sincerity.* Chicago: University of Chicago Press.

Khan, Faris. 2016. "Khwaja Sira: Dissent, Sex/Gender Activism, and State Regulation in Pakistan." Paper presented at the Feminist Preconference: Sexuality, Gender Identity, and Sedition, 45th Annual Conference on South Asia, University of Wisconsin, Madison, October 26.

Krell, Elias. 2013. "Contours through Covers: Voice and Affect in the Music of Lucas Silveira." *Journal of Popular Music Studies* 25, no. 4: 476–503.

Madrid, Alejandro L. 2009. "Why Music and Performance Studies? Why Now? An Introduction to the Special Issue." *TRANS: Revista Transcultural de Música* 13. Accessed October 12, 2016. http://www.sibetrans.com/trans/articulo/1/why-music-and-performance-studies-why-now-an-introduction-to-the-special-issue.

Muñoz, José Esteban. 1999. *Disidentifications: Queers of Color and the Performance of Politics.* Minneapolis: University of Minnesota Press.

Prasad, Pavithra, and Jeff Roy. 2017. "Ethnomusicology and Performance Studies: Towards Interdisciplinary Futures of Indian Classical Music." *Musicultures* 44, no. 1: 187–209.

Reddy, Gayatri. 2005. *With Respect to Sex: Negotiating Hijra Identity in South India.* Chicago: University of Chicago Press.

Rivera-Servera, Ramón. 2009. "Musical Trans(actions): Intersections in Reggaetón." *TRANS: Revista Transcultural de Música* 13. Accessed October 2, 2016. http://www.sibetrans.com/trans/article/62/musical-trans-actions-intersections-in-reggaeton.

Roy, Jeff. 2015. "Ethnomusicology of the Closet: (Con)Figuring Transgender-*Hijra* Identity through Documentary Filmmaking." PhD diss., University of California, Los Angeles.

Roy, Jeff. 2016. "Translating Hījṛā into Transgender: Performance and Pehchān in Mumbai's Trans-Hījṛā Communities." *Transgender Studies Quarterly* 3, no. 3–4: 412–32.

Roy, Jeff. 2017. "From *Jalsah* to *Jalsā*: Music, Identity, and (Gender) Transitioning at a *Hijra* Rite of Initiation." *Ethnomusicology* 61, no. 3: 389–418.

Soneji, Davesh. 2011. *Unfinished Gestures: Devadasis, Memory, and Modernity in South India.* Chicago: University of Chicago Press.

Sugarman, Jane. 1997. *Engendering Song: Singing and Subjectivity at Prespa Albanian Weddings.* Chicago: University of Chicago Press.

Waugh, Thomas. 2001. "Queer Bollywood, or 'I'm the player, you're the naive one': Patterns of Sexual Subversion in Recent Indian Popular Cinema." In *Keyframes: Popular Cinema and Cultural Studies*, ed. Matthew Tinkcom and Amy Villarejo, 280–302. New York: Routledge.

Weidman, Amanda. 2006. *Singing the Classical, Voicing the Modern: The Postcolonial Politics of Music in South India.* Durham, NC: Duke University Press.

# PART III

# THE POLITICS OF SOUND

.....

# 8

# Banlieue Sounds, or, The Right to Exist

*Hervé Tchumkam*

.....

This analysis of the relation between *banlieue* sounds and the right to exist builds on the critical contention that the so-called global South is not simply a spatial designation: it also exists within the North. In this chapter, I am concerned with Afro-descendent French citizens living in France, which is not typically considered a space of the global South. My approach to sound builds not on sound theory *sensu stricto* but on oppositional couples such as visibility and invisibility and audibility and inaudibility. I not only suggest that, precisely because of its former colonial relation with Africa and its current treatment of its citizens of African descent, France (or the French banlieues, at least) has gradually taken the form of a global South space.[1] I also examine the recent civil disobedience and rioting in the *cités*, those "projects" on the outskirts of cities where African migrants and their offspring live. I argue that this civil disobedience can be interpreted as a response to the apathy of the French government in the face of the peaceful March for Equality and Against Racism of 1983, as well as successive governments' refusal to hear and listen to French hip-hop singers and rappers who have been using their voices to denounce social injustice for the past three decades.[2] In fact, in November 2005 France was inundated by waves of violence in the cités. While these uprisings were viewed as having particular political significance, it should be noted that rioting as a way to denounce social injustice via the performance of one's existence has been a permanent fixture of the banlieues for thirty-five years and was employed as recently as the summer of 2013 in Trappes.

The reaction of several French governments to such social unrest has essentially been to draw a dividing line between "them" and "us"—that is, between those deemed unworthy of accessing (or unable to access) Frenchness, at least as it is constructed in official discourse, and those capable of respecting the values of the Republic. Along these lines, a powerful discourse more or less condemning the cités as *zones de non droit* (lawless zones) has gained acceptance, while Parliament has simultaneously attempted to pass a law to glorify French "colonial grandeur" in their former colonies. For the French government and media, as well as for some intellectuals, the banlieues represent a threat to Frenchness, a cradle for thugs and barbarians whose aim is to take over France. In this chapter, I challenge the idea of endemic violence that affects young people of African descent in the marginalized peripheries of France. Using a contrapuntal approach, my argument is twofold: first, building on Didier Lapeyronnie's concept of "primitive rebellion," I argue that urban riots are a way for marginalized social groups to claim their right to exist; and second, I turn to other forms of insurrection and suggest that the ultimate form of youth resistance in the French banlieues might be characterized by what Giorgio Agamben (1993) has called the coming community. I start by looking at what I call the "colonial continuum" from colonial Africa to contemporary France to justify my theoretical grasp of the French banlieues as a space whose inhabitants are as much part of the global South as are those who live in the favelas in Brazil, the townships of apartheid South Africa, the ghettoes of the United States, and the slums of Cameroon.

## THE COLONIAL CONTINUUM: FROM AFRICAN MIGRATIONS TO FRANCE AND THE BANLIEUES

To better understand what I call the colonial continuum, it is necessary to return to the fundamental problematic of immigration and national identity in contemporary France. What does it mean to be an "alien" or "foreigner" in France today? This question is particularly relevant to the understanding of social unrest in France, because when we talk about the banlieues, we are talking less about immigrants than about their offspring who were born in France and represent the third, and sometimes fourth, generation of heirs to immigration. Citizenship laws have existed in France since 1889. At that time, however, immigrants were not coming from French colonies, and further, they were not "visible" minorities—that is, they did not exhibit a bodily difference, such as skin color. A significant element of French imperialism in

*Hervé Tchumkam*

Africa, the Caribbean, and Asia, therefore, was political behavior that did not hear racial Others and thus rendered them nonexistent.

It is worth underlining that, in the late nineteenth century, France defined foreigners (*étrangers*) as those who were not citizens: Jews were regarded as foreigners from within France, for example, while migrants from other European countries were foreigners from outside. Strikingly enough, it was Napoleon Bonaparte and his followers who considered Jews foreigners in France, as would be made only too clear during the Dreyfus affair at the turn of the twentieth century; this was evident again in the Vichy era during World War II.[3]

In 1894, the year that the Dreyfus affair began, a certain Capitaine Danrit published *L'invasion noire* (1894), a novel that tells the story of France being invaded by a horde of bestial-sounding black Islamist warriors led by the Turkish sultan.[4] In the novel, the warriors are stopped at the gates of Paris by a lethal gas, which is used to prevent them from taking over the last standing cradle of civilization in the European world. The novel, which would have been considered pure fantasy by contemporary readers, can be regarded as a fictional signal that Frenchness and civilization versus barbarianism would become a long-standing fight. The people from Africa who are represented in Danrit's novel as threats to liberty and civilization would eventually constitute the third category of foreigners in France—that is, non-Western natives, or Africans. African people were French not in terms of citizenship but as colonial subjects; they were French in that they had duties (e.g., fighting for France in the war and paying taxes) but no rights (Cooper 2009).[5]

In 1906–1907, France encouraged economic migration. In addition, colonial subjects were called to fight alongside France in World War I. This, combined with successive waves of labor migration from Africa, left France stuck with its colonial empire within its own national borders. To put it bluntly, the colony was no longer only overseas—henceforth, its logic, and organization, would exist within France itself. By the early twentieth century, the colony had been deterritorialized, and France was witnessing a transformation in the notion of the foreigner: the "étranger" was no longer a European migrant but an indigenous person or native. In other words, the *postcolonial* foreigner was born. And since that time, postcolonial immigration has been particularly fraught, because the postcolonial subject has become undesirable, unwanted. In the minds of French people, the fundamental question is now, "Can the postcolonial Other coming from the French colonial empire, with a different culture and religion, be assimilated into France and share national space with French people?" It might have made sense to ask this question when the first generation of African migrant workers and soldiers arrived in France in the early twentieth century; today, however, in the case of the banlieues, what is being called into question is not the "Frenchness" of those who migrated from Africa but that of their children and grandchildren who were born in France.

Long before the urban riots in 2005, but even more so since then, France, the so-called country of human rights, has become the site of all kinds of debates and controversies over national identity. Even a cursory review of newspapers; television and radio commentary; political speeches; and talk on the streets, in the schools, at coffee shops, and in stadiums reveals this. And the debate about national identity is characterized by two main factors. First, it stigmatizes African migrants and their offspring as "immigrants" (as if a Spanish or Italian person who migrates to France is not also an immigrant). And, second, it perpetuates France's systematic erasure of the past, an oblivion that is particularly interesting in that it emphasizes respect for the laws of the Republic, as well as the need to protect France from invasion by barbarians and the rise of *communautarisme* (the idea that people of African descent in France "ghettoize" themselves instead of trying to mingle with the rest of society), to use a word that is often employed in the rhetoric of segregation in contemporary France. We are thus facing what the sociologists Didier and Éric Fassin (2006) have called the "ethno-racialization" of social relations in France, in which people of African descent are locked into the segregated prisons of hopelessness that constitute the French banlieues. The notion of citizenship is replaced with that of origin; descent replaces the function of the passport that one does or does not hold. To put it more bluntly, not being white is now considered a major transgression in France, if not a crime, and one of its repercussions is that nonwhite French citizens are reduced to survival, and their mode of existence is reduced to systematic control, humiliation, scorn, and discrimination.

Neither religion nor country of origin—or even the issue of citizenship—justifies how the French political powers have treated the cités over the past four decades. The situation in the banlieues can be explained via what I call the "colonial obsession," with the colony now located within the metropolis. France's attachment to its colonies seems so powerful that, more than fifty years after independence, the descendants of former colonial subjects (even those born in France) continue to be regarded as bodies at the disposal of that nation. The treatment of the banlieues invites us to rethink what Aimé Césaire said in *Discourse on Colonialism* ([1955] 2000)—namely, that the West has entered a cycle of self-*decivilization*. Césaire argues that, with regard to

*Hervé Tchumkam*

colonization, Western civilization is a decadent civilization. He also makes a statement that may have been shocking at the time it was written but that seems truer than ever when one looks at the history of colonization, Franç-afrique, and the French banlieues:[6] "[what the twentieth-century Christian bourgeois] cannot forgive Hitler for is not *the crime* in itself, the *crime against man*, it is not the *humiliation of man as such*, it is the crime against the white man, and the fact that he applied European colonialists procedures which until then had been reserved exclusively for the Arabs of Algeria, the 'coolies' of India and the 'niggers' of Africa" (Césaire ([1955] 2000: 36).

Not only have the banlieues become privileged sites for the duplication of colonial rules, but they also stand out as areas that, although contained within the state, represent the most feared threat to that very state. Thus, the cités are subjected to exclusion from within—what Agamben (1998: 8) calls "inclusive exclusion"—and young people have next to no hope for better social conditions. If one looks back at the French state's reaction to the uprisings of November 2005, one is left with the impression that, at the levels of urban distribution of space and use of sovereign violence, the French cités and banlieues have become an extension of the unequal international relations between France and its former colonies on the African continent.

Not only are the French banlieues gradually taking the form of ghettoes,[7] as Lapeyronnie (2008) has brilliantly shown, but they also call for historical work on the colonial and the postcolonial. Such work is still largely perceived in academic circles as representing a risk to French "national" cohesion and as tending to weaken the Republic, as became evident in the debates that surrounded the law on colonialism of February 23, 2005, which, among other things, required high-school teachers to frame French colonialism in positive terms (Bancel 2009: 168). Such is the context that fuels frustration and ignites uprisings in the banlieues. After all, as recently as March 17, 2011, Claude Guéant, then the French interior minister, uttered words in an interview with the French newspaper *Le Figaro* that I deem highly suggestive of what may have been the feelings of young people of African descent in the cités during the years and months preceding the wave of social unrests in France since 2005: "A government must pay attention to what is necessary and should be attentive to the demands of the people. . . . The French people feel that uncontrolled immigration has modified their landscape. They are not xenophobic. They want France to remain France."[8] Not only does such a statement read as a blueprint for exclusion, but it also suggests that the banlieues, the majority of whose inhabitants (*banlieusards*) are French citizens of African descent, are not "French." Youth resistance culture in the form of urban riots, therefore, appears to some to be the only way to claim their right to exist in a society in which they are silenced and ignored.

In France, violence has become the ultimate response to the political status quo, inducing a new form of community that I elucidate later. First, I examine urban riots as modes of existence and modes of speech, taking the reception of sounds — or their absence — as hermeneutic points of departure and arrival. Although French rap and hip-hop artists have been producing antiestablishment sounds since the early 1980s, as noted earlier, they have been largely ignored by the French government. However, when in 1995 the French hip-hop group Suprême NTM called for rebellion in the song "Qu'est-ce qu'on attend?" (What Are We Waiting For?) — rapping, "You have asked for a war of the worlds, here it is / But what, but what are we waiting to light the fire? / But what are we waiting for before we stop following the rules of the game?" — the call did not go unnoticed.

In the case of the French banlieues, the issue is no longer whether the subaltern can speak. It is, rather, whether the sounds and speeches that the subaltern produces are heard or simply dismissed. Hence, there is a need for subalterns to find other ways to make their voices heard and to have their claims taken seriously, especially if one follows Laurent Dubreuil's (2012: 101) remark, with regard to education in the colonies, that "the most striking and characteristic methods of French colonial usage are: a tendency toward the confiscation of the speech of the colonized, a grammatical hypercorrectness that aims to deny the colonized's ability to speak *properly*, and a constant annihilation or denial of the colonized *parole*. Thus, the culmination of colonial censorship is silence, deafness and promotion of *parlure*." What stands out in this quote is that, having been already disqualified from the realm of using language and delivering correct or appropriate speeches, silence is imposed on the colonized — and on their descendants in France — as a way to prevent them from raising their voices and being heard.

What is at stake here is less speech itself than its content, which, in the cases of both colonial Africa and the French banlieues, takes the form of demands for emancipation and social justice. Hence, precisely because they are controlled by the dominant French political apparatus, all of the speeches, novels, and songs are made, if not nonexistent, at least irrelevant to hear. In such a context, it only makes sense that the reinvention or the appropriation of speech by banlieue youths adopts the form of a scream (*cri*). While Dubreuil (2012: 102) claims that "the subaltern (or any comparable role), when she speaks, is especially at risk of continuing a *parlure* that has been made for her to 'stay still,'" I differ slightly and argue that, precisely because of its

*Hervé Tchumkam*

unpredictability, the new forms of sound to which the rioters resort—such as that of cars burning and Molotov cocktails exploding—escape the trap of being conditioned, sanctioned, and thus predestined to silence.

## WHEN FLAMES REPLACE SOUNDS: ON URBAN RIOTS
## AND YOUTH RESISTANCE

Riots and unrest in the banlieues have become a response to the apathy of French political leaders and the silence of state power—a silence that implies an active politics of denying recognition. Consider the polemic between the French philosopher Raphaël Enthoven and the French rapper Kery James in 2015, after James released a song in memory of Zyed Benna and Bouna Traoré, the two teenagers whose deaths sparked the riots in 2005. The polemic is particularly revealing of the ways in which the intelligentsia, like the mainstream media and the French political elite, silence Afro-descendent French citizens who are claiming their right to exist as French citizens amid the debate on national identity that was launched in France in 2009. Moreover, the fact that, ten years after the riots, a rapper's song calling attention to the lack of an indictment of the police officers who were present when the teenagers died and denouncing the stigmatization and racism would be met with a response reducing him to his skin color suggests not only the inefficacy of acoustic explorations of life but also the irrelevance of such sounds, which is coterminous with the irrelevance of the life of the community from which the singers come. How do we make sense theoretically of such a dead end?

"Behind the mask of reason and civility," Achille Mbembe (2009: 48) writes, "every old culture conceals a nocturnal face, a vast store of obscure drives that, given the opportunity, can turn lethal. . . . In the West that nocturnal face and those drives have always been fixated on race, the Beast whose existence the French Republic, in its blind concern for universality, has always refused to admit." Taking my cue from Mbembe, I argue that it is clear that the colonial fracture still has strong repercussions in French society. That fracture can be summarized by the fact that, on the one hand, there are people of African descent who are thirsty for knowledge about their history and, foremost, for an understanding of the past that informs the present and enables a future that can be dreamed about or planned. On the other hand, there is the French government that has managed consistently to refuse confronting the realities of colonization and has even failed to acknowledge the pain that it has caused in many, if not all, of its former colonies, particularly in Africa. Rather, trying an impressive *coup de force* to pass laws such as that of February 23, 2005, to institutionalize the "glorious past" of France and the "positive effects" of colonization has been the main concern of the French government. Unlike the "cunning of recognition" that Elizabeth Povinelli (2002) describes in Australia, France does not recognize past atrocities committed by the state; nor does it acknowledge the horror of these actions. Instead, it is adamant about reinvigorating what it views as the colonial nation's past grandeur and its core values. Moreover, the media are famous for printing distorted stories of the riots and for censoring and silencing voices from the periphery. The banlieues are thus caught in a paradoxical position between the rhetorical feat of a state that proclaims it upholds human rights and a daily life in which these very rights are permanently trampled.

Officially, the inhabitants of the cités are painted as the "scum" of society, which allows for the further oppression of the already oppressed. How can one be a French citizen and yet be treated as a second-class citizen? How do invisible minorities in French society move to visibility? How does one stop being an "outsider within"? The French sociologist Ahmed Boubeker (2009) has analyzed the situation of visible minorities, who are invisible not so much because they cannot be seen as because those with political power ignore their presence. Drawing on Ralph Ellison's *Invisible Man* (1952), a novel that highlights the relation between race and sound, Boubeker argues that rioting—or, at the very least, audibility—can serve as a channel to obtain self-affirmation and recognition. Offering a very powerful insight into the situation of the banlieues, he shows how the paradox of visibility and invisibility reveals the failure of French universalism. I argue that the paradox of visibility and invisibility overlaps with the problematic of existence and nonbeing.

The people in the banlieues, like Ellison's invisible man, are invisible merely because their compatriots refuse to see them; they do, in fact, exist, but their existence is unacknowledged and challenged by the French political apparatus. In response to their rejection by the state, they have created their own modes and codes of existence. Patrick Alessandrin's film *Banlieue 13—Ultimatum* (2009) offers a perfect illustration of how the banlieusards make life livable even in the face of rejection and extreme state violence. One character in the film claims that, because the state had ghettoized them, the banlieusards have become more powerful and united by creating their own society within French society, in which resilience has become a modus operandi. Violence and uprising in the marginalized French suburbs thus has become a mode of existence, with words of protest and the sounds of hip-hop and rap being replaced by flames rising from cars that have been set ablaze. It is only in the moments of confrontation between young people and the

*Hervé Tchumkam*

police that verbal interaction seems possible, because in daily life, it is the police who speak and do not allow young people even to explain themselves. Yet youth violence itself is used by the state and the media to exacerbate the construction of the banlieusards as monsters—similar to the figure of the African native who needed to be brought from the darkness of barbarianism into Enlightenment. The fate of banlieusards fighting for their right to exist and that of the nationalist fighters in Africa battling for independence is also striking: both rely heavily on violence for self-affirmation, and just as voiced demands prompted colonial violence in Algeria and Cameroon, they gave rise to sovereign police brutality in the banlieues. While the historical circumstances of colonization (i.e., the civilizing mission) are used to justify colonial violence, it is very difficult to understand objectively why, "with no place of recognition to hang their hats, these new barbarians in the news do not even have the extenuating circumstance of being the offspring of generations of poverty and oppression" (Boubeker 2009: 77).

From this standpoint, I contend that the cités have become a privileged site for Franco-African international relations—or, better, a site where the logic behind the nebulous idea of Françafrique is applied. I base my thesis essentially on the argument that racial hierarchy, the leading principle that surrounded the idea of the civilizing mission, also seems to be the principle that is guiding the government's reactions to urban violence in the France's secluded and marginalized peripheries. To put it bluntly, more than a half-century after the wave of independence of former French colonies, France's sovereign power seems to have located an extension of the colony that needs civilizing. The statement by Interior Minister Guéant quoted earlier was not unique: on January 7, 1990, Michel Rocard, then France's prime minister, said, ironically referencing Frantz Fanon, "We cannot welcome all the wretchedness of the world" (quoted in Boubeker 2009: 75). A few years later, French President Jacques Chirac voiced his pride in what many perceive as the glorious past of France, saying, "Pacification, development of the territories, the spread of education, the establishment of modern medical practices, and the creation of administrative and legal institutions are all marks of that indisputable work to which the French presence contributed, not only in northern Africa but also on every continent" (quoted in Bancel 2009).

But the urban riots of 2005 introduced a new dimension to the history of social unrest in France—namely, what Lapeyronnie has called "rioting as collective action," which refers to the peculiar way in which these riots spread throughout almost the entire country. Building on the postulate that rioting is collective, Lapeyronnie emphasizes that, contrary to media depictions and official state rhetoric, rioting belongs to the normal repertoire of political action; thus, violence, vandalism, and looting are "ordinary behaviors." Youth resistance stands out as a response to injustice, for "the discrimination is exacerbated by the general impression that the police enjoy impunity, which, in the eyes of the young, allows officers to strip them of all rights and exert unrestricted power over them" (Lapeyronnie 2009: 30). Other testimony from rioters (Muchielli 2006) also attest to the fact that frustration is so high among young people with no hope that rioting has become the only language at their disposal, even though, during my field research (which was prompted by my own experience as a black man who has lived and studied in France), most of the young people acknowledged that the violence can be self-destructive.

As Mehdi, one of the young people I spoke with while conducting research in the French banlieues, put it, "What else have we got to lose? We were born dead, and we are condemned to rot behind these bars, with nobody willing to listen to our voices." Mehdi's statement signals that the limit, or threshold, between life and death, audibility and inaudibility, is no longer relevant: the boundaries have been blurred by a social system in which he oscillates between living and dying, or the limit between life and death has been erased to the point that, like Mehdi, many young banlieusards feel they have been born into a world in which they are already dead. Just like Agamben's figure of the *homo sacer*, that citizen of ancient Rome who could not be ritually put to death but whose death was not considered homicide, the young people in the banlieues have simply become true bodies of exception.

Here again, the analogy between Franco-African relations and the handling of the cités by French authorities points to a disturbing analogy between Africa and the banlieues in the eyes of the French government. In both colonial and postcolonial Africa, and in the banlieues, the angels of law and order have become the apostles of disorder, and any pretext related to human rights is welcome to victimize Africans on the continent, as well as young people of African descent living outside the walls of major French cities. Everything seems to indicate that the specter of the Algerian War, as well as that of African nationalists in Cameroon (see Deltombe et al. 2011)—and, in some ways, the black revolutionaries of Haiti—still hover over France. In any case, the relegation of the banlieues to some sort of dustbin of the Republic evokes the perception of Africa in the colonial imaginary as it was brilliantly developed in Albert Memmi's *The Colonizer and the Colonized* (1957).

In the French suburbs, however, the heirs to colonization and immigration are choosing to shift the position of the oppressed. Youth resistance culture in the banlieues hence offers interesting paths for understanding the culture

*Hervé Tchumkam*

of the oppressed. They have decided not to allow humiliation and scorn to go unanswered. Breaking the paradox of the invisibility of a visible minority and the chains of silence requires that the young people in French ghettoes claim their right to exist. In doing so, they riot to form a collective "us," in opposition to an antagonistic "them" consisting of scholars and pundits who have "turned the tables on [them] by putting forth arguments that range from the breakdown of African families living in France, to polygamy, rap music, and more generally their unwillingness or perhaps inability to assimilate as more accurate explanations for their marginalization" (Gondola 2009: 147).

Similarly, rioting is temporarily replacing rap and hip-hop as a form of resistance to discrimination; as such, it exemplifies a refusal of the place the young rioters have been assigned in French society. During the rebellion in the French banlieues, it has not been unusual to hear rioters scream, "You've been deaf to our cries and insensitive to our situation. Now you're going to listen to us!" while charging the police, breaking down the doors to public buildings, or shattering the glass at bus stops. The sound used during rebellion is not only very violent; it also communicates anger, frustration, despair, and consciousness of irrelevance in the eyes of the dominant political system. For many observers of French society, the youth violence amid the social unrest of 2005 dramatized in the allegory Pierre Morel depicts in *Banlieue 13* (2004), a futuristic action movie about alleged plans by the French government to destroy the banlieues with a bomb intentionally put into the hands of a famous drug lord. It is always tempting to draw a clear dividing line between fiction and reality, between real life and the imaginary, but one might wonder whether the future of the banlieues will be better or worse that the one depicted in *Banlieue 13*. Such a question, prompted by Morel's film, becomes even more vivid when one pays attention to Hacène Belmessous's book *Opérations banlieues* (2010).[9] In all, the rejection of the political system—as a world to which they have a right but that marginalizes and rejects them—takes the form of extreme violence. In the words of Lapeyronnie (2009: 41):

> The use of violence in rioting turns out to be both a perfectly rational strategy of exerting pressure and the result of a lack of autonomous agency, of a strong dependence on a system to which they are denied access. Rioters are thus defined both by their marginality and by their dependence. They are "outsiders," victims of a system that rejects them, discriminates against them, and ultimately "prevents them from living." Yet they are also the "poor," the "insignificant," who feel heavily dependent, especially in political terms, on this same system and this same society.

Paul Silverstein and Chantal Tetreault (2006) reject the media discourse that associates upheavals in the French suburbs with gang violence. The violence in the banlieues is, instead, an exhibition (or demonstration) of the coming together of singularities that assert their being and claim their place in the distribution of the sensible—that is, the violence is ultimately about a group of young French citizens who oppose being reduced to their bodies, following the conception of immigrants as having value only in terms of their potential for capital production based on the bodies of their offspring. The more France sucks the labor out of black and Arab French citizens, the more radiant and powerful the reduction of people to their bodies becomes. Take, for example, the controversial soccer star Zinedine Zidane and his teammate Lilian Thuram. Despite their achievements, which added to the glory of France, they have remained what the philosopher Mohamed Sidi Barkat (2005) calls *corps d'exception*—bodies that are dangerous and unworthy of being regarded as citizens. They are thus reduced to mere organic and dehumanized bodies.

## UNREST IN THE FRENCH BANLIEUES AND THE AFRICAN RENAISSANCE

If youth resistance culture in France is a claim for the right to exist, in what capacity can this battle benefit Africa, the continent from which most of the rioters' (grand)parents came and to which the rioters themselves seem to be limited by the dominant discourse? How can any drive from impoverished and secluded French ghettoes provoke changes in the configuration of Françafrique? Or, put quite simply, what can Africa learn from the uprising and youth resistance in the French banlieues?

On the relation between the treatment of Africa by France and the treatment of its cités, Isabelle Coutant (2005) has shown that young people who hail from the French suburbs refuse to be treated as their kin were in colonial times; at the same time, she points to the economic exploitation that prevents African countries from developing. For residents of the cités, the perpetuation of the colonial past can no longer be tolerated. Mbembe (2010: 94) presents a similar analogy between colonial rule and the treatment of the banlieues, offering an interpretation that reaches back to the era of slavery:

> The most important scene of the Republic's brutality and discrimination was [the] plantation during slavery and the colony beginning in the nineteenth century. Quite directly, the problem raised by the

*Hervé Tchumkam*

plantation system and the colonial order is that of the functionality of race, which becomes the principle for the exercise of power and, by the same token, the condition of sociability. In today's context, to talk about race is to call for a reflexion about difference, about he or she with whom one does not share anything or just a little—about those whose presence, albeit with us, near us or among us, are ultimately not ours. Before the Empire, the *plantation* and colony were elsewhere and entailed strangeness and distance—something overseas. The extreme limits of those territories continue to shape their presence in metropolitan imaginaries. Nowadays, the plantation and the colony have been displaced here, outside the walls of the city (in the banlieue).

On sovereign violence in African colonies and the uprising in 2005, Silverstein and Tetreault reflected on the cités through the theoretical lens of what they call "postcolonial urban apartheid." For them, the "colonial law's deployment in response to the present crisis points to an enduring logic of colonial rule within postcolonial metropolitan France. Like settler cities of the colonial period, contemporary French urban centers function in opposition to their impoverished peripheries, the latter being consistently presented in the media, state policy, and popular talk as culturally, if not racially, different from mainstream France" (Silverstein and Tetreault 2006).

The troubling situation and lack of horizon for Afro-descendent French citizens in the French suburbs and the colonial logic operating in former French colonies share features on least at two levels. The first level is the Othering of blacks and Arabs; and the second involves the "polemic distribution of spaces, times, and forms of activity that determines the very manner in which something in common lends itself to distribution" (Rancière 2004: 12). The Othering of blacks and Arabs has to do with the ways in which the Afro-descendent French citizens, because of their origins, are reduced to silence in the same way their grandparents were in the colonies. And continuing this comparison between the colonies and the banlieues, the spatial division of French cities into center and periphery is also reminiscent of the division between the white administrative neighborhoods and natives' slums of colonial times. In both cases, one is faced with a context in which the sounds exuding from peaceful calls for justice became inaudible. The spatial partition in the colonies and in today's banlieues thus also share a blurring of the threshold between audible and inaudible—or, in other words, the clouding of the limit between existence and nothingness.

Lapeyronnie insists that youth violence in France is not an outlaw act but, rather, a clearly definable political gesture. At the same time, however, he argues that the rioters are "primitive rebels" because their actions cannot be reduced to a list of categorized demands. Primitive rebellion as destructive yet rational violence severs reality and brutally displays the existence of the rioters, however, in ways that *cannot* be expressed as a list of demands. This close relationship between unidentifiable demands and public manifestations in Lapeyronnie's thought certainly echoes Boubeker's (2009: 87) conclusion that, "where the official data see only integration problems, anomie, the spleen of ghetto, or the stigma of exclusion, we need to rediscover the living subject, the swarming mass of humanity that eludes categorization by little republican schemas."

Hence, while France is faced with the absence of clear demands from the rioters, the Republic has proved incapable of categorizing young people performing their right to exist through violent outburst. It is precisely at the intersection of these two thoughts, I propose, that Africa can derive lessons about modifying its status vis-à-vis its relation to France. When I speak of Africa here, I am not referring to "official Africa," represented by heads of state who for the most part — especially in French-speaking Africa — are puppets of the West. Rather, I am suggesting that a renaissance in Africa can be inspired by popular resistance by youth in the French suburbs. As Didier Gondola (2009: 164) notes, "There can be no doubt that the swelling number of the African diaspora and the status of second-rate citizens that indexes most Africans in France is attributable, albeit only partly, to the minorization of Africa." Gondola goes even further, stating that, with regard especially to French-speaking Africa, the renaissance of the continent "hinges to a large extent on the ability of 'Black France' to decolonize the République and respond to [Simon] Gikandi's call that the 'task of decolonization must be taken *to the metropolis itself*; the imperial mythology has to be confronted on its home ground'" (164, emphasis added).

In other words, youth violence and urban riots in France are sound testimony that, following a reversal of situation, change is occurring in the belly of the beast: it is now the French crusaders for national identity who have become the beasts while the young rioters display their right to exist. Boubeker, Lapeyronnie, and Gondola all seem to be suggesting that the upheavals of 2005 in the banlieues have unveiled the worst enemy of the French state. That enemy is not the barbarian Muslim horde of Capitaine Danrit's *L'invasion noire*; Nicolas Sarkozy's "scum'"; or even the twenty-first-century terrorist feared by most "civilized" nations, but the "coming community"

(Agamben 1993) made up of "primitive rebels" and "outsiders within" in the French banlieues.

By constituting themselves as an unidentifiable, unclassifiable, and unpredictable group, young people of African descent in the cités of France have shown that the path to asserting the right to exist is youth cultural resistance. In so doing, they have also succeeded in shaking the Republic's certainties about the power of life and death in the banlieues and in the (post)colony, thereby hinting at the weakness of the imperial project. As Étienne Balibar put it in a recent interview, what is most important about the uprisings in the French banlieues in the fall of 2005, is "to observe that the language of the revolt was quite different from what most politicians and pundits had announced both inside and outside France. It had very little to do with multiculturalism and religion, or rather the dimensions of the social conflict were incorporated into a more general discourse on justice and dignity, what I call *droit de cité*, a right of both citizenship and allowance, and therefore of residency" (Balibar 2009: 323). Will French-speaking Africa be able or willing to learn from the French banlieues and break the chains of silence, affirm its right to exist, and claim its place in the distribution of the sensible? While that question remains to be carefully investigated, it is worth signaling that, ultimately, the banlieue sounds I analyze in these pages, beyond empirical sounds of cars exploding or objects clashing during upheavals are first and foremost an act of speech that is carved in actions. In this sense, it is clear that not only are the riots a direct and more efficient way to gain visibility, but they also stand out as the ultimate disqualification of the silence that is imposed by state power on banlieue sounds that are meant to be heard. Clearly, banlieue sounds in rioting, well beyond sounds of music and interviews, are actions that spoke for youths and forced their visibility on French society, no matter how much France pretends not to see them or not to listen to them.

*Notes*

1. I use the term *banlieues* here to refer to the projects that are populated mainly by Afro-descendent French citizens and African migrants and that are generally situated at the outskirts of major French cities.

2. For further elaboration of the link between the march in 1983 and the rioters of 2005, see Beaud and Masclet 2006.

3. Alfred Dreyfus was a French officer of Jewish descent who had been accused of treason and sentenced for allegedly revealing French military secrets to the Germans, only to be exonerated a decade later.

4. The name "Capitaine Danrit" is believed to be an anagram for Émile Augustin Cyprien Driant (1855–1916), a French Army officer. The attention garnered by the Dreyfus affair may account for why the publication of *L'invasion noire* went largely unnoticed.

5. For an elaborate discussion of citizenship and colonial subjects in French history, see Cooper 2009.

6. For a detailed account of how what was considered a mere friendly cultural relationship between France and its former African colonies turned into the neocolonial plundering of Africa by France, see *Françafrique: 50 années sous le sceau du secret* (2010), directed by Patrick Benquet, Compagnie des Phares et Balises, Paris. Also, interesting developments of the issue are in Verschave 2003, 2005. For a more recent perspective, see Pesnot 2011.

7. It should be pointed out that, looking at historical process, the American ghettoes—unlike the French banlieues, which are the result of colonization—were constructed according to a logic based on transatlantic slavery. For an elaboration of the French banlieues as ghettoes, see Lapeyronnie 2008.

8. "Immigration: Claude Guéant provoque l'ire des socialistes," *Le Figaro*, March 17, 2011, accessed March 17, 2011. http://www.lefigaro.fr/politique/2011/03/17/01002 -20110317ARTFIG00637-immigration-claude-gueant-provoque-l-ire-des-socialistes .php. All translations from French are mine.

9. In the book, Belmessous (2010) provides the reader with an extensive investigation on the preparedness of the French Army, Special Forces, and police to intervene in the banlieues as in an urban warfare zone. The thesis of the book is that the banlieues have become a threat to national identity in France that the government has decided to eliminate. The book abounds with confidential documents and testimonies of first-rank French officers. In sum, according to Belmessous, terrestrial military engagement on the banlieues is no longer taboo and in some ways it has been already authorized should the government decide to give it clearance.

### References

Agamben, Giorgio. 1993. *The Coming Community*, trans. Michael Hardt. Minneapolis: University of Minnesota Press.

Agamben, Giorgio. 1998. *Homo Sacer: Sovereign Power and Bare Life*, trans. Daniel Heller-Roazen. Stanford, CA: Stanford University Press.

Balibar, Étienne. 2009. "Interview." In *Communities of Sense: Rethinking Aesthetics and Politics*, ed. Beth Hinderliter, William Kaizen, Vered Maimon, Jaleh Mansoor, and Seth McCormick, 317–36. Durham, NC: Duke University Press.

Bancel, Nicolas. 2009. "The Law of February 23, 2005: The Uses Made of the Revival of France's 'Colonial Grandeur.'" In *Frenchness and the African Diaspora*, ed. Charles Tshimanga, Didier Gondola, and Peter Bloom, 167–83. Bloomington: Indiana University Press.

Barkat, Sidi Mohamed. 2005. *Le corps d'exception*. Paris: Éditions Amsterdam.

Beaud, Stéphane, and Olivier Masclet. 2006. "Des 'marcheurs' de 1983 aux 'émeutiers' de 2005: Deux générations sociales d'enfants d'immigrés." *Annales*, no. 4: 809–43.

Belmessous, Hacène. 2010. *Opérations banlieues: Comment l'état prépare la guerre urbaine dans les cités françaises*. Paris: La Découverte.

Boubeker, Ahmed. 2009. "Outsiders in the French Melting Pot: The Public Construction of Invisibility for Visible Minorities." In *Frenchness and the African Diaspora*, ed. Charles Tshimanga, Didier Gondola, and Peter J. Bloom, 70–88. Bloomington: Indiana University Press.

Césaire, Aimé. [1955] 2000. *Discourse on Colonialism*, trans. Joan Pinkham. New York: Monthly Review.

Cooper, Frederick. 2009. "From Imperial Inclusion to Republican Exclusion? France's Ambiguous Postwar Trajectory." In *Frenchness and the African Diaspora*, ed. Charles Tshimanga, Didier Gondola and Peter J. Bloom, 91–119. Bloomington: Indiana University Press.

Coutant, Isabelle. 2005. *Délit de jeunesse: La justice face aux quartiers*. Paris: La Découverte.

Danrit, Capitaine [Émile Augustin Cyprien Driant]. 1894. *L'invasion noire*. Paris: Flammarion.

Deltombe, Thomas, Manuel Domergue, and Jacob Tatsitsa. 2011. *Kamerun! Une guerre cachée aux origines de la Françafrique 1948-1971*. Paris: La Découverte.

Dubreuil, Laurent. 2012. "Notes towards a Poetics of Banlieue." *Parallax* 18, no. 3: 98–109

Fassin, Didier, and Éric Fassin. 2006. *De la question sociale à la question raciale?* Paris: La Découverte.

Ellison, Ralph. 1952. *Invisible Man*. New York : Random House.

Gondola, Didier. 2009. "Transcient Citizens: The Othering and Indigenization of Blacks and Beurs within the French Republic." In *Frenchness and the African Diaspora*, ed. Charles Tshimanga, Didier Gondola and Peter J. Bloom, 146–66. Bloomington: Indiana University Press.

Lapeyronnie, Didier. 2008. *Ghetto urbain*. Paris: Robert Laffont.

Lapeyronnie, Didier. 2009. "Primitive Rebellion in the French *Banlieues*: On the Fall 2005 Riots." In *Frenchness and the African Diaspora*, ed. Charles Tshimanga, Didier Gondola, and Peter J. Bloom, 22–46. Bloomington: Indiana University Press.

Mbembe, Achille. 2009. "The Republic and Its Beast: On the Riots in the French *Banlieues*." In *Frenchness and the African Diaspora*, ed. Charles Tshimanaga, Didier Gondola, and Peter J. Bloom, 47–54. Bloomington: Indiana University Press.

Mbembe Achille. 2010. *Sortir de la grande nuit*. Paris: La Découverte.

Memmi, Albert. [1957] 1991. *The Colonizer and the Colonized*. Boston: Beacon Press.

Muchielli, Laurent. 2006. "Les émeutes de novembre 2005: Les raisons de la colère." In *Quand les banlieues brûlent*, ed. Laurent Mucchielli and Véronique le Goaziou, 11–35. Paris: La Découverte.

Pesnot, Patrick. 2011. *Les dessous de la Françafrique: Les dossiers secrets de monsieur X.* Paris: Nouveau Monde.

Povinelli, Elizabeth A. 2002. *The Cunning of Recognition: Indigenous Alterities and the Making of Australian Multiculturalism.* Durham, NC: Duke University Press.

Rancière, Jacques. 2004. *The Politics of Aesthetics*, trans. Gabriel Rockhill. London: Continuum.

Silverstein, Paul A., and Chantal Tetreault. 2006. "Postcolonial Urban Apartheid." *Social Science Research Council.* Accessed June 11, 2016. http://riotsfrance.ssrc.org /Silverstein_Tetreault.

Verschave, François-Xavier. 2003. *Françafrique: Le plus long scandale de la République.* Paris: Stock.

Verschave, François-Xavier. 2005. *De la Françafrique à la mafiafrique.* Paris: Tribord.

*Hervé Tchumkam*

9

# Sound Studies, Difference, and

# Global Concept History

*Jim Sykes*

.....

## ON ASTROLOGY AND SONIC PROTECTION

In 2007, I was on a brief hiatus from fieldwork in Sri Lanka and visiting
New York when a friend told me that drummers were needed for a unique
upcoming event. The legendary Japanese band Boredoms was looking for
seventy-seven drummers to play in a concert to be held on the waterfront at
Brooklyn Bridge Park. Far from an ordinary drum circle, the event—called
77 Boadrum—would include seventy-seven drummers playing full drum
kits arranged in a coil (like a boa constrictor), performing for seventy-seven
minutes on July 7, 2007 (i.e., 7/7/07). Yamataka Eye, the leader of Boredoms,
would play a seven-necked guitar constructed for the occasion.

The genius of the event was its simplicity. The band arranged their own
music, which we drummers had not heard in advance (save for a few drummers marked as "section leaders," shown in white circles in figure 9.1). The only
major instruction was to copy whatever the drummer to your right was doing.
During the concert, the members of Boredoms were positioned at the center of
the boa on a stage (numbered 0–3 in figure 9.1; Eye did not drum and was given
a zero); drum patterns that began at the center spiraled out slowly, as one drummer after another copied the pattern played by the drummer to the right. Both
for the sheer physicality of sitting amid that many drum kits played all at once
and because we were playing a well-organized composition but had no idea
how it would unfold, 77 Boadrum was a memorable experience for all involved.[1]

The number seven, of course, has spiritual significance for many cultures.
Eye claims he got the idea for 77 Boadrum while visiting the ancient Mayan

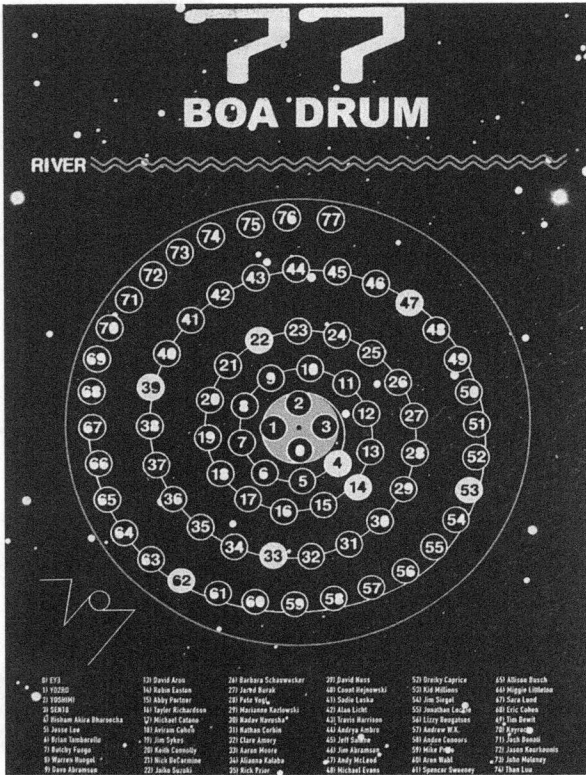

FIG 9.1.
The layout for
77 Boadrum.
Photograph by
the author.

ruins at Palenque, where he noticed that the Sun Temple had seventy-seven steps. In an interview, he explains how he understands the significance of seven and its relations to the boa and sun:

> The number seven . . . has a form that expresses an energy that spins clockwise: it represents a system of movement, like the energy in a DNA helix. It's all related. Actually, the "bore" in "Boredoms" means "boa constrictor." A boa expresses an energy from the earth. This simple, long shape increases its energy by spinning and turning around. It's something fundamentally connected to the number seven and also to the sun. Looking for inspiration for 77 Boadrum, I'd go to a river a bit north of our house. I'd see a snake. The third time I saw one it was all coiled up and looking at the morning sun.[2]

About a week after the concert, I was back in Sri Lanka, interviewing my friend Rohana Wasantha about Sinhala Buddhist approaches to sound.[3]

Wasantha is a devout Buddhist and practicing astrologer who performs rituals at *devales* (shrines for deities), structures that are often housed on the grounds of Buddhist temples. He told me that when certain sounds are produced by people engaged in right mind and right action (the paradigmatic example being the chanting of the Buddha's teachings, or Dhamma, in the Pali language by Buddhist monks), the result is like an electric charge, akin to "a firewall that protects your computer from unwanted viruses." Wasantha stressed that it is not sound alone that has this power but certain kinds of sound, produced by certain kinds of people (defined according to their karma and intentions), at specific times, facing in certain directions. I immediately thought of 77 Boadrum and told Wasantha about the concert.

A few weeks later, I met up with Wasantha again, and he had a gleam in his eye. He had done the astrological chart for 77 Boadrum and written up the results for a Sinhala astrology journal. I have provided the chart here (see table 9.1), as well as a translation of his comments on the event.[4]

After presenting the astrological chart, Wasantha interprets his findings:

> According to this horoscope and its planetary positions, we can give a better vision of the juncture. The date was Saturday; it represents the 7th day of the week. This date is controlled by Saturn. Saturn represents hardness, difficulties, sadness, sorrow and also tiredness. This (from 6:36 PM to 7:36 PM) Hora lord is Mercury.[5] During the middle 12 minutes of the Hora, poison power [is] generated. This means that the poison period starts at the 24th minute of the Hora. *That is 07:00 o'clock PM*, the time duration controlled by Venus. Why is it that we call this "the poisoning time"? Saturn and Venus are enemies; as they shine, they emit a poisoning power to the universe, *as does the sound with it.* (emphasis added)

Wasantha then explains the purpose of 77 Boadrum. It turns out we were doing more than just entertaining an audience:

> This Natal chart [table 9.1] shows the Zodiac sign as Scorpio. Its lord is Mars. Mars is placed in the sixth house in this horoscope. The horoscope's sixth house represents the enemy and diseases. Mars represents the chief leader of the army. Now you can guess the prediction of this event.
>
> The United States suffered a terrorist attack on September 11th. It caused much damage to the country and the state. *This 77 Boadrumming functions to give protection to the state.* See the Lg. that placed

TABLE 9.1 Natal Chart

| | |
|---|---|
| Date: | July 7, 2007 |
| Time: | 19:00:00 |
| Time Zone: | 4:00:00 (West of GMT) |
| Place: | 75 W 10′ 15″, 42 N 20′ 39″, Brooklyn, New York, USA |
| Altitude: | 0.00 meters |
| Lunar Yr-Mo: | Sarva-jit, Nija Jyeshtha |
| Tithi: | Krishna Ashtami (Ra) (22.05% left) |
| Vedic Weekday: | Saturday (Sa) |
| Nakshatra: | Revati (Bu) (38.66% left) |
| Yoga: | Atiganda (Ch) (27.48% left) |
| Karana: | Kaulava (Ma) (44.11% left) |
| Hora Lord: | Chandra (5 min sign: Dhanu) |
| Mahakala Hora: | Sukra (5 min sign: Kanya) |
| Kaala Lord: | Mangala (Mahakala: Mangala) |
| Sunrise: | 5:36:21 |
| Sunset: | 20:34:32 |
| Janma Ghatis: | 33.4860 |
| Ayanamsa: | 23-57-48.86 |
| Sidereal Time: | 13:01:27 |

Jupiter in the first house. Jupiter has a friendship with the first house because its Zodiac sign at that time was Scorpio. Jupiter gives advice, prosperity and wealth. This planetary position is more favorable to the state also. But Sun and Mercury are placed in the eighth house in this horoscope. It shows the low dignity of the state at the time.[6] Jupiter is in conjunction with Saturn and the Moon in a triangular vision. All three of these planets are in the major Triangle of the horoscope. *This combination gives destroying power, but in a diplomatic way.* The educational sector will be fostered by the state; and so will priests according to this combination. . . . This shows [that] musicology is engaged with wisdom. (emphasis added)

Wasantha's astrological reading determined that 77 Boadrum thwarted a terrorist attack on the U.S. government. His paper then concludes by noting that the spiral position of the drummers enhanced the protective power of the sound:

*Jim Sykes*

When observing these planetary positions in conjunction with the spiral organization of the 77 Boadrum, we can *see* the power that was produced by this show's sound. It started from the center and spiraled outward. This shows that the spiritual hardness transmits [a] bass sound to the universe. But it gives protection to the state. Also, drummers receive protection through the power of sound [and] the velocity of its bass rhythm, and they will gain fame worldwide. (emphasis added)

## WHAT THIS CHAPTER IS ABOUT

I am hardly the first researcher to feel the anthropological gaze turned back on himself or herself by a smart interlocutor. Wasantha's astrology paper deftly situates (remaps?) the 77 Boadrum concert according to a Sinhala Buddhist ontology of sound as protection that I explore throughout this chapter—a conceptualization that aligns, in many respects, with Eye's own conception of the event. This is a sonic ontology in which the positions of planets, the timing of sound, and the spatial orientation of sound producers generates sonic power that protects the sovereign while also protecting the population at large. In what follows, I argue that the idea of sonic protection has a long history and contemporary political relevance in Sri Lanka. More broadly, though, I make the case that certain sounds produced by Sinhala Buddhists gain protective efficacy through sonic gift exchanges between human and nonhuman beings such as gods, demons (*yakku*), tree spirits, and nonhuman animals. Humans are just one kind of being that is able to "unlock" and "wield" the power of sound; such sounds can be vigorously guarded or passed on, creating connected sonic histories across various levels of being. Thus, Sinhala Buddhist sacred sounds are not merely "devotional," if we take that word to mean the self's outward expression of an inward, private emotion; as Theravada Buddhists, Sinhalas do not believe that the self exists in an objectively real, stable, and persisting way, as is captured by the Christian notion of the soul. Rather, sounds in Sinhala Buddhist religious contexts are best understood as objects that are separate from the self, which gain their meaning through their (frequently *public*) exchanges with nonhumans and their positioning in relation to stars, gods, objects, and so on. One aim of this chapter, in sketching out this sonic ontology, is to "remap" sound and the concepts of sound producer and sonic efficacy in multiple ways, much as Wasantha did in his analysis of 77 Boadrum.

I want to stress, though, that my overall aim here is to go beyond simply marking out an alternative sonic ontology and placing it under the banner of sound studies. As Gavin Steingo and I state in the introduction to this volume, we do not think the project of remapping sound studies should simply be one of cordoning off the global South and marking it as a place for sonic difference.[7] Nor do I intend merely to repeat the argument made artfully by Charles Hirschkind (2006) that *contra* the ocular-centrism of Western modernity, sound is embodied and the relations between sound and body are interpreted differently in different locations and traditions (in Hirschkind's study, for Muslims in Cairo).[8] Rather, my principal goal is to use a discussion of the persistence of Sinhala Buddhist sonic ontology in postcolonial Sri Lanka to lay the groundwork for a retheorization of the historical development of "religious sounds" in modernity, one that rethinks their relations to global historical processes such as colonial-era political liberalism, Christian missionization, and postcolonial ethno-nationalism. Analyses of "sound" in what Andrew Sartori (2008) calls "global concept history," I argue, should not simply tell a story of transformation in which the historical development of "sound" as a concept in the West (which I consider in detail in the next section) can be mapped onto the rest of the world as the proper story about the development of sound in global sonic modernity. By contrast, I show in this essay that non–Judeo-Christians have strategically appropriated, ignored, incorporated, or rejected the Western-derived secular notion of sacred sound (as disenchanted and as intimately related to the expression of an emotional state) depending on whether it was useful to them in any given context.[9] This means recognizing that the spread of media technologies does not necessarily isolate sound from the other senses but may be used to enhance preexisting sonic ontologies in which sound is connected to stars, gods, demons, malignant supernatural glances, and so on.[10]

The point I really want to drive home, though (and that emerges at the end of this essay), is that colonial-era Christian missionaries misunderstood Sinhala Buddhist religious sounds because they thought those sounds were understandable simply in isolation, though ultimately defined as the outward expression of an inward emotional state (the soul). I fear that there is an obsession—even a fetishization—in sound studies with sound "as such," because it is believed that favoring sound is necessary to overturn the ocular-centrism of modernity. But modernity everywhere throughout the world is not ocularcentric, and the emphasis on sound "as such" risks promoting ideas about sound that in the global South have a colonial Christian heritage—one that did damage in Sri Lanka through missionary denigrations of Sinhala religion.

*Jim Sykes*

## SECULARISM, SOUND STUDIES, AND
## CONNECTED HISTORIES

Despite the diversity of its subject matter, sound studies has produced a surprisingly consistent narrative on the development of "sound" as a concept
in Western modernity. The perspective emerges forthrightly in Jonathan
Sterne's *The Audible Past* (2003), but it can be found in many other places besides: Leigh Eric Schmidt's *Hearing Things* (2000); Charles Hirschkind's *The
Ethical Soundscape* (2006); Veit Erlmann's *Reason and Resonance* (2010); and,
more recently, Ana María Ochoa Gautier's *Aurality* (2014) and Isaac Weiner's
*Religion Out Loud* (2013), among others. At the risk of oversimplification,
all of these scholars explore how the privileging of vision became integral
to Western modernity through the generation of a set of binaries between
sound and vision (which Sterne famously called the "audiovisual litany"), in
which sound, speech, and orality became associated with the word of God,
while writing and vision became associated with scientific progress and secular modernity. Hearing and speech were idealized as "manifesting a kind of
pure interiority" (Sterne 2003: 13)—that is, the Christian notion of the soul—
so the distinction between sound and vision is "essentially a restatement of
the longstanding spirit/letter distinction in Christian spiritualism" (Sterne
2003: 16).[11] The secularization of sound in the West thus developed, paradoxically, through sound's conceptual separation from vision as the sense most
attuned to (Christian notions of) the spirit, interiors, and immersion. While
each of these scholars documents the effects of the litany inside and outside
North America and Europe in her or his own ways, none calls for a return to
orality as a return to God. For example, Sterne (2003: 16–17) bases his text on a
rejection of that call as it was made earlier by the Jesuit scholar Walter Ong.[12]
Thus, sound studies has operated through a secular, social constructivist
perspective that documents the historical growth of the ideological position
it presumes.[13]

Within this framework, writings on religion in sound studies have tended
to emphasize what Hirschkind calls "the relationship between technologies
of auditory discipline and the historical redefinition and repositioning of
religion [as] central to the emergence of secular modernity" (Hirschkind,
in Sterne 2012: 64; Eisenlohr 2011). For example, Alain Corbin's *Village Bells*
(1998) shows how an auditory technology (church bells in nineteenth-
century France) helped generate modern, secular notions of time; Shane
White and Graham White's *Sounds of Slavery* (2006) documents how sounds

on North American plantations functioned to interiorize African religious practices while furthering capitalism by regulating labor. Weiner's *Religion Out Loud* argues that noise regulation laws in the twentieth-century United States have acted to protect the supposedly shared, secular nature of public space but in practice have often discriminated against the sounds of minority religions (e.g., the Islamic call to prayer). All this is to say that not only the foundation of sound studies but much subsequent work in the discipline shares a kinship with a strain of thinking about secularism (associated predominantly with Talal Asad [2003], Saba Mahmood [2015], and Charles Taylor [2007]), and in which a Protestant Christian vocabulary and agenda are shown to have generated much of "secular" modernity. Through the "Protestant secularism" of the law, written on behalf of the need to regulate capitalist labor, sacred sounds were deemed rightfully to belong to the private domain of the household and places of worship (outside the workweek), defined as "devotional" practices that are also outward productions of the soul (i.e., the identity) of a community. To put this as straightforwardly as I can, "secularism" is *not* "not religious"; rather, it is a configuration of public and private space in which public space is seen as naturally disenchanted and private space as the rightful place for the expression of religious belief and practice. This makes religious sounds in public space appear as the emergence (and possibly the transgression) by the community and its beliefs in gods and ghosts and penance (and so forth) *into* a public space that is rightfully defined by the workings of capital—that is, monetary exchange. This ignores that some sacred sounds in their first instance are defined as *public* forms of exchange (e.g., with gods [Sykes 2015]).[14]

The famous distinction Asad made was between the political doctrine of secularism, which strives to keep religion out of politics, is central to definitions of modernity, and is currently "hegemonic as a *political goal*," and "the secular," which is conceptually historically prior to "secularism" and "a domain of historically constituted and variably related behaviors, sensibilities" (Agrama 2012: 7). Modernity, Asad (1993: 36) writes, "requires not the production of a uniform culture throughout the world, but certain shared modalities of legal-moral behaviour, forms of national-political structuration, and rhythms of progressive historicity." Asad understands the secular as a transcultural phenomenon, for it has been spread far and wide through culturally specific modes of entanglement (encounters between formations of the secular and indigenous ways of being) that can be studied. As Agrama (2012: 1) writes, "We no longer see the domains of the religious and the secular as given, but rather, as mutually constitutive of each other in often tense and contradictory ways." Mahmood (2015: 3) puts it this way: "Secularism . . . is not simply the organizing structure for what are regularly taken to be a priori elements of social organization—public, private, political, religious—but a discursive operation of power that generates these very spheres, establishes their boundaries, and suffuses them with content, such that they come to acquire a natural quality for those living within its terms."

This means that we must consider the imagined isolated listener walking down the street listening to her iPod (an important figure in earlier work in sound studies) as a formation of the secular that discursively produces the secular when we universalize her and turn her into an icon of sound studies.[15] And so, too, for the voluminous recent writings in sound studies that avoid religion and the secular altogether while discussing sound in global cities, sound in everyday life, and the spread of sound through digital technology, the implication being that religion and the secular are auxiliary topics that can be subgenres of sound studies rather than understood as processes that generate(d) our very definitions of "sound," "the city," "self," and "technology" in the first place (as Sterne, Hirschkind, and some of the other scholars mentioned earlier emphasize). As a formation of the secular itself, the normative methodology of sound studies—to look for how sound reproduction technologies increasingly isolated sound in modernity—is a historically emergent Western position that should be projected onto others only with much care, hesitation, nuance, and reflection. Sure, the question of how the audiovisual litany has been produced in the non-West should be an important part of sound studies, but pursuing that question should not foreclose the possibility that the spread of the litany was not totalizing or (in some places) all that successful. More to my point, since the contrast between religion and the secular is a false binary "produced posthoc by the ideological lens through which the Western present views the past and elsewhere as premodern" (Canell 2010: 91; see also Asad 1993, 2003)—much like enchantment/disenchantment and nature/culture—to assume that the story of sound in global modernity is ipso facto a story about the spread of the West's audiovisual litany is to assume that sacred sounds in the non-West have been thoroughly Christianized and that those that persist according to their original terminologies and values are necessarily unmodern. The search for sound "as such," in other words, defines Christian approaches to sound and its relations to the self as normative, modern, and related to technological development and progress.

Scholarship that documents the effects of the audiovisual litany outside the West has been beneficial for at least two reasons. First, sound has now been incorporated into what had been understood simply as the ocular-centrism inherent in the production of colonial knowledge and power — what Ochoa Gautier (2014: 13) calls Western modernity's "despatialized omni-science." And second, sound now emerges as relevant to studies (embodied by the writings of Partha Chatterjee [1986], Ritu Birla [2009], Andrew Sartori [2008], and others) that demonstrate how, through laws and attitudes in the British colonies, political liberalism (the premier philosophy of nineteenth-century Britain) conceptually located culture, religion, and community in a feminized private sphere and politics and economics in a masculinized public sphere. Colonial liberal laws, in other words, were the modus operandi through which sacred sounds were interiorized (secularized).

As far as Sinhala Buddhists are concerned, there are three major sets of rituals performed largely by a caste called the Beravā ("drummer"): *deva tovils* that ask gods for protection from natural disasters such as drought and pestilence; *yak tovils* or rituals that heal individuals suffering from diseases brought on by demons (*yakku*); and *bali tovil*, a single ritual of offerings to planetary deities that wards off malignant planetary influences. Sinhala processions (*peraheras*) often include these ritualists but involve a broad swathe of the community and are held in association with prominent Buddhist temples on full moon nights. All of these practices involve a lot of sound, particularly drummers, although *tovils* rely heavily on *mantras* (magic spells) and sung poetry. They are, in their original construal, public events of political importance (since they generate social hierarchies in public spaces and constitute communal decisions about maintaining the wellness of society) and economies (in that they involve offering sounds to gods or demons in return for protection from calamity or the healing of illness). But as Michael Roberts (1995) has shown, in colonial Ceylon the government issued a number of noise ordinances and a system of procession licenses that regulated the sounds of religious processions, which became defined legally as "noise" when they appeared to threaten public order (Sykes 2017). Colonial law became a way for populations to jockey for position as they filed legal complaints about the noise of others to gain the upper hand for their practices. Another aspect of the liberal transformation of Sinhala society during this period was the replacement of Sinhala divisions of time and the workweek with Western ones (though for holidays, the lunar calendar is still used [Roberts 1995]). Religious practices that used to be tacked onto one another (thus making an event go on for several days straight) took an all-night form, occurring basically from sunset to the early hours of morning.

The next phase in the story, in which the conceptual, spatial, and temporal transformations of religious practices facilitated by political liberalism made them amenable to anticolonial and postcolonial nationalism, is better known. In Sri Lanka, it facilitated the growth of a staged tradition in which Beravā dance and drumming practices were institutionalized in schools for the arts (rather than schools for Ayurvedic medicine) and made open to all castes to learn. A healing ritual (the Kohomba Kankariya) performed by the Beravā caste came to serve as the foundation for Kandyan Dance (Sri Lanka's national dance), resulting in what had been an all-male ritual tradition being dominated largely by female dancers doing staged routines (Reed 2010). The process began as early as 1919, when Kandyan dancers were inserted into the annual Buddhist procession the Äsala Perahera (now the country's biggest tourist event), helping turn it into an embodiment of "national culture" (Ambos 2011: 255). Another transformation was the changes wrought to *dharmadesana*, a term generally translated as "sermon." This was achieved through the efforts of late nineteenth-century Buddhist reformers such as the lay Buddhist revivalist Anagarika Dharmapala (1864–1933), through which a ritualistic tradition of reciting Pali texts whose power lay in large part on their generating merit for the listener through "the evocational experience of the sheer sound of the text" (Bate 2005: 472) became transformed so that its value came to reside more in its meaning and the monk's explanation of it (Deegalle 2006; Seneviratne 1999: 49–50).[16] Finally, Western medicine stigmatized certain Ayurvedic medicinal practices that involve harnessing supernatural power through sound, particularly the large-scale healing rituals called yak tovils (which many branded as exorcisms [Scott 1994]).

It would thus seem that Sinhala Buddhist sounds were transformed by rationalization and communalization stemming from political liberalism and its counterpart, "Protestant secularism," under the "despatialized omniscience" of colonial rule and the needs of the postcolonial state (for "traditional culture"). But here I want to argue that after the Anglo-educated leaders of the first postcolonial generation passed away, they were replaced (by the late 1970s) with a new generation of leaders, many of whom hailed from the rural south and did not have the Anglo education of their forebears and who invested heavily in efficacious ritual (Kapferer 1997). Sonic protection, I suggest, became a core component of Sinhala Buddhist nationalism, even seeing a resurgence during the civil war (1983–2009). For much of the Sinhala population, the audiovisual litany never took hold, *even though* the discourses promoted by colonialism and early postcolonial nationalism (i.e., the idea of a "national dance") became second nature. Older notions of sonic efficacy and exchange were not jettisoned, but new meanings were layered on top of or intersected with them.

## BUDDHISM, NOT-SELVES, AND THE EFFICACY
## OF EXTERIORIZED SPEECH

> When Basil Rajapakse [the president's brother] visited the site of the mine explosion, *a pirit* [Buddhist chant] CD was still playing in the vehicle of this once feared politician. . . . There were also two large parcels of cashew nuts in the jeep and seeing this Basil had said, "He always brings cashew nuts for *aiya* [brother] and me."[17]

This passage from *The Island* newspaper describes the aftermath of the assassination of D. M. Dassanayake, Sri Lanka's minister of nation building, in a roadside attack on January 8, 2008 (reportedly by the Tamil rebel group the Liberation Tigers of Tamil Eelam). Dassanayake was known for "playing dirty" in politics, such as the incident during the Wayamba provincial elections in 1998 when he and his colleagues stripped political rivals naked in broad daylight and chased them down the street, throwing stones.[18] The use of food in Basil Rajapakse's recounting of the event is meant to convey Dassanayake's humanity: he gave cashew nuts to his friends. The fact that Dassanayake was listening to Buddhist chant (*pirit*) at the time of his death demonstrates that he had "cooled off" in recent years, becoming more reflective and spiritual (in a "rationalized" sense). Mentioning that the pirit CD continued to play while death and destruction reigned all around conveys that the Sinhala Buddhist nation will survive terrorist attacks. But because having one's mind focused on the Triple Gem (the Buddha; his teachings, or Dhamma; and the community of monks, or Sangha) at the time of death is thought to guarantee a good rebirth, Rajapakse's statement is also meant to reassure the public that, despite his bad behavior in life, Dassanayake is certain to achieve a good rebirth.

Pirit still carries much "protective" potency in contemporary Sri Lanka (the word *pirit* means "protection"). When monks chant, they hold a thread (*pirit nul*) that may touch an object, such as a copper plate. Recitation channels protective power into the thread, charging the object. People buy copper plates that have been charged this way and hang them on their walls to protect their homes. The copper plate is an acousmatic trace (Kane 2014) in that it is a physical reminder (and *remainder*) of the power of past sound (but not a reminder of sound *as such* but rather that the Buddha's teachings were sounded by monks at specific times, repeated an auspicious number of times, while touching a thread that touched the plate). Wasantha told me he took part in long pirit ceremonies, lasting up to six months or more, in which monks switched off and kept the recitation going until the text was repeated the appropriate number of times. The Sinhala ontology of sound as protection has its roots, perhaps, in the story about the beginning of pirit (Pali, *paritta*) reported in a famous text in the Pali canon, the Ratana Sutta. The Buddha received pleas from the inhabitants of the city of Vesali to help save them from disease (*rogabhaya*), demons (*amanushahbaya*, or "fear of nonhumans"), and starvation (*durbikshabhaya*); he asks his disciple Ananda to go to the site to recite the Dhamma. When Ananda did so, the city was saved. Today, pirit is used for both serious and mundane acts of protection — for example, to protect the national cricket team before important matches.

Pirit is just one component of a broader everyday world of Sinhala speech as protection in which speech gains power when it is linked with objects and images. A mantra is a "magic spell" that contains combinations of words drawn from a mixture of languages. The mantra gains power when it has life breathed into it (*jīvam kirīma* [Scott 1994: 219]), which involves repeating it an auspicious number of times (e.g., 108, 21, or 7 times) "so that the energy or vibration . . . of the words imparted through uttering them strikes the smoke of the *dummala* or resin as it rises from the fire pan" (Scott 1994: 219). Mantras may gain power when they are recited alongside geometric patterns called *yantra*, a combination of sound and image similar to what Wasantha found in 77 Boadrum. Similarly, the science of architecture, *vastu*, requires positioning rooms and statues of the Buddha and deities in relation to one another; new homeowners may invite a member of the drummer caste (the Beravā) to drum in the direction that the god of death (Yama) is residing, another way to protect the house. Before the foundations of a house are built, yantra may be buried at its four corners to protect the house, but they "have to be buried at a moment of perfect silence," for if the *yantra* are "caught" by the voice of a human or animal, it renders their protection invalid (Argenti-Pillen 2002: 95).

In her ethnography on how trauma from the war and second Janatha Vimukthi Peramuna (JVP) uprising (a Marxist uprising of Sinhala youths in the late 1980s) was experienced by Sinhala villagers in a slum in southern Sri Lanka, Alex Argenti-Pillen (2002: 85–101) describes how certain sounds create illness and require acts of "acoustic cleansing" to dispel them. She marks a division in Sinhala village society between the *gedera* (house) and *samādjaya*, or wider society; some sounds from the outside world (which she calls "the wild"), such as that of insurgents wandering through neighborhoods shouting insults, drunks stumbling home at night, and bombs and murders, generate terrified hearts (*hita bayayi*) and illness. The walls of the house are a barrier from the gaze of the wild but are sonically porous; hence, villagers perceived a "non-Buddhist wild society encroaching on the ethos of the household" (Argenti-Pillen 2002: 87). On top of this, talk *about* disease can produce illness, as can "the poisonous voice," which occurs when the jealous words of another manifest in one's saliva and are "swallowed," leading one to ingest the faults "inherent in the other's jealousy and anger" (Argenti-Pillen 2002: 91). Consequently, villagers tend to talk around illness rather than address it directly.[19]

One way Sinhala villagers may dispel sonic illnesses, Argenti-Pillen (2002: 92) claims, is by "telling it to the tree": a person goes into "the wild," approaches a milk or jak tree, tells the problem/illness to the tree (thus "tying" it to the tree), and then dissipates the negative energy by cutting the tree with a knife. If an illness is due to the negative influence of demons (*yakku*), a large-scale public ritual (*yak tovil*) may be necessary. The ritual ties the malignant glance (*dishtiya*) of the *yakkha* to an object (or ritualist), after which it is cut by reciting mantras (Scott 1994). While drumming is important to yak tovils, it may arouse the "gaze of the wild" and thus can also make people fall ill, allowing it to be used as "acoustic revenge" against one's neighbors.[20] Elsewhere (Sykes 2018), I document how yak tovils helped confront trauma from the Indian Ocean tsunami of 2004, as when a young girl talking in a garbled voice was found to be possessed by her dead aunt, speaking as though she had water in her lungs. Following the tsunami, events were held (such as one I attended at the famous Galle Face Hotel) in which excerpts of yak tovils were staged as "traditional culture" to raise money for those who suffered from the tsunami—a de-efficacized healing ritual held for the purpose of healing. While this might seem to reinforce the common view that investments in astrology, sorcery, yak tovils, and so on are a rural phenomenon, this is not entirely true. In 2015, for example, then-President Mahinda Rajapakse infamously called (and lost) an early election due to advice from his astrologer (who, as it happens, is also a former director of the National Savings Bank). When I returned to Sri Lanka in 2015, my drum teacher (who also identifies as an astrologer) told me he had just been hired by a major corporation to recite mantras to drive a ghost out of the basement of a large building in Sri Lanka's capital, Colombo.

In the late 1950s, Sri Lanka was in political turmoil after a wave of Sinhala Buddhist nationalist sentiment and polices, particularly the "Sinhala Only" law of 1956 that made Sinhala the only language of government, which

*Jim Sykes*

had left the island's minority groups (particularly the Tamils, who make up 18 percent of the population) disenfranchised. In 1959, Prime Minister S. W. R. D. Bandaranaike was assassinated by a Buddhist monk. This period is usually regarded as having laid the groundwork for the civil war (1983–2009), in which Tamil rebel groups fought for an independent homeland in the island's north and east. At the time, some observers blamed the first line of the country's newly adopted national anthem, "Namo Namo Matha" (written by the musician Ananda Samarakoon to a melody that may have come from the Indian nationalist and Nobel Laureate Rabindranath Tagore), which had an inauspicious arrangement of syllables, for the country's political problems. In Sinhala Buddhist thought, patterns of three syllables can be auspicious or inauspicious depending on the ordering of their long and short syllables; the patterns, called *ganachandas*, have their roots in Pali chant. The first line and title of the national anthem were changed, but Samarakoon (who suffered from depression) committed suicide. It bears mentioning that many Sri Lankans found the idea of blaming *ganachandas* for the country's political problems ridiculous, but that is precisely my point: both perspectives are present in contemporary Sri Lanka. Sometimes they are at odds, while in other contexts (such as in the Dassanayake incident), a "rational" investment in sound may be compatible with an investment in sonic efficacy.

In his second sermon, the Buddha propounded the doctrine of "not-self" (*anattā-vāda*), the idea that "beings have no soul, no abiding essence" (Gombrich 2006: 47). Buddhists challenge any "static, unalterable dogma that posits a permanent and reincarnating self or person" (Collins [1982] 1990: 76). While talk about the self is acceptable in everyday speech, there is a "refusal to speak of a self or permanent person in any theoretical contexts" (Collins [1982] 1990: 78). For Buddhists, a person consists of five *skandhas* (aggregates, heaps), including "the body, feelings, perceptions, impulses and consciousness," and "the belief in a self or soul, over these five *skandhas*, is illusory and the cause of suffering" (Humphreys 2012: 51). I suggest that the lack of a permanent self or soul makes Sinhala Buddhist sounds easily conceptualized as exterior to the "self" (or, rather, "not-self"), and that by doing so, sound easily functions as a form of exchange between humans and nonhumans.

To build a drum, the Beravā caste traditionally make offerings (including drumming and sung poetry) to spirits that inhabit the tree to be chopped down for the drum. Another example is the performance of drums and shawm (*horanāva*) in Buddhist temple ceremonies, which is called *sabda pujāva*, or "sound offering." A ritual I saw in the southern low country, the Bera Poya Hevisi, presents the history of Beravā drumming as a story of sonic gift exchange. On the day of the Buddha's Enlightenment, the gods played music that had been taught to them by *ghandarvas*, lower-rung Buddhist deities; these lines of drum poetry (*padas*) were given by Pulastya (son of Brahma and grandfather of Ravana, the "evil king" in the Ramayana epic) to Sri Lanka's indigenous population, the Väddas. The Väddas gave them to the Beravā, and the Beravā give them back to the gods in the Bera Poya Hevisi.[21]

All-night rituals for deities (*deva tovils*) are offerings (of sound, incense, dance, food, and so on) that ask gods to protect villages from natural disasters such as drought and pestilence. Premakumara de Silva (2000: 46) notes that these rituals were held during Sri Lanka's war as nationalistic "cultural displays" (*sandharsana*) that were *also* efficacious rituals to protect soldiers fighting in the war zone. At the University of the Visual and Performing Arts in Colombo, young women and men training to be dance teachers perform excerpts of ritual dances at outdoor concerts for their parents; in one such concert I attended, after the students' performance concluded (around midnight), the students were replaced by Beravā ritualists who continued until daybreak. Even though the event began as a staged performance of a traditional ritual, not completing the ritual could render the event inauspicious.

The political relevance of sonic protection in contemporary Sri Lanka operates through a Buddhist "zoopolitics" that contrasts with the Western version explored by Ochoa Gautier (2014). In Sinhala parlance, the Buddha deemed the island a place for the protection of Buddhism, as Dhammadipa (the island of the Buddha's Dharma), and thus "to fight for Sri Lanka is to fight for Buddhism, and vice versa" (Bass 2013: 47; Bartholomeusz 2002). While European zoopolitics uses a discourse on culture, the lettered city, and the arts to demarcate who falls on the side of culture and who falls on the side of a less-than-human animal nature (Ochoa Gautier 2014), for Buddhists, non-human animals have always been conceived as past and future humans and future Buddhas and thus in a sense can be on the side of culture rather than nature. For example, a case of animal preaching is found in the *Sasa Jataka* (a story about a past life of the Buddha); a group of animals are confronted with an ethical conundrum that one animal solves, serving as an example to the others. The animal reveals himself to be the Bodhisatta (the Buddha-to-be). The issue of animal listening is pursued by James Stewart (2017: 58), who suggests that Sinhala Buddhists do not think animals can understand the Dhamma, but "hearing the Dhamma, being proximate to dhammic institutions, or even merely witnessing representatives of the Dhamma, can lead to a better spiritual condition." The incursion of European zoopolitics into Sri Lanka may have marked a distinction between nature and culture,

*Jim Sykes*

but I suggest this was mapped onto, without eliminating, indigenous divisions between beings (conceived through the concept of karma). This is a zoopolitics where animals may find themselves on the way toward culture if they hear the sound of the Buddha's teachings, while humans may find themselves slipping toward a nonhuman, animal-like nature if they fail to listen to them. The act of offering sound to gods to protect soldiers was premised, perhaps, on claims of a different ontological status for the non-Buddhists against whom the soldiers fought.

One highlight of the Independence Day celebrations of 2010 (the first to be held after the end of the war) was that they included a performance of the Kohomba Kankariya, the ritual (mentioned earlier) that serves as the foundation for Sri Lanka's national dance, Kandyan Dance. The festivities, Eva Ambos (2011) notes, positioned the president in a "lineage tradition," since it conceptualized him as falling in the lineage of the precolonial Kandyan kings. The ritual rhetorically situated the Sinhalas back in the times described in the ritual, which recalls events that supposedly happened a few decades after the progenital ancestor of the Sinhalas, Vijaya, arrived from North India in the sixth century BCE. The event might have been an ethno-nationalist display promoting Sinhala homogeneity in the public sphere, but it also "reinforced [the] image of the king and the unitary nature of his kingdom" (Gamage 2007), a practice with roots in precolonial Sri Lanka. When greeting visiting dignitaries, drummers and dancers have long facilitated *dākum*, the ritual recognition of the king's sovereignty, which in the colonial period needed to be done to obtain trading rights (Roberts 2004). A similar sort of dākum occurred in 2015 when drummers and dancers flanked Pope Francis and Sri Lanka's President Maithripala Sirisena when the pope visited the island.

## COLONISTS, MISSIONARIES, AND THE AUDIO-VISUAL LITANY

Look at the decay of the temple and disregard of the sacred places, as evidence of the decay of the faith. Look at the horrible devil ceremonies and cruel superstitions to which the people have been driven by the atheism of Buddha; and you will form a more correct idea of what Buddhism really means than by reading the old Buddhist fairy tales transformed into beautiful English poetry. (Langdon 1886: 53)

Nineteenth-century British colonists and missionaries (and Sinhala Buddhists!) were by no means unified on what constitutes proper Buddhism.[22] In general terms, Elizabeth Harris suggests that nirvana was considered negatively as annihilation, and Buddhism was deemed atheistic, but Buddhist ethics were considered "in essence, benign and praiseworthy" (Harris 2006: 48).[23] Positive appraisals were reserved for concepts such as meditation and enlightenment and psychological terms such as *dukkha* (suffering or unsatisfactoriness [Harris 2006: 48]).

While all Buddhist texts and practices were deemed atheistic, I suggest that the domain of indigenous ritual (conceptualized as orality/speech) was particularly associated with demon worshiping, while writing/representation (i.e., the Bible and the "benign" ethics in Buddhist texts) were associated with the word of God and the possibility of scientific progress and indigenous morality, respectively. This is perhaps only subtly presented in the passage by Langdon quoted earlier, but the argument is supported by the ubiquity of printings of the Bible by missionaries in colonial Ceylon and the fact that missionaries were the first to make Sinhala- and Tamil-language dictionaries. The lettered city, in other words, was not fully secular. Writing/representation was not only on the side of secular modernity but also about bringing orality/speech and indigenous religion into the lettered, cultured, colonial, Christian domain, though some (like Langdon) were frustrated because Buddhism seemed to be founded on an atheism so severe it rendered Sinhalas apathetic, making them hard to convert (Scott 1994: 165). This is to say not that the voice as the word of God did not play a role for Christian missionaries (sermonization remained key) but, rather, that the letter was also important for conversion, and the sounds of Sinhala rituals were placed on the side of orality/uncultured/animal nature ("horrible devil ceremonies") and as something altogether different from Buddhist texts (which seemed to hold Buddhist "ethics" — though both were deemed atheistic).[24]

The key here is that the project of missionization required sublimating the Sinhala ontology of sound as protection because it invoked beings (gods, demons) that missionaries wanted to eradicate and that they viewed as not really a part of Buddhism. Beravā rituals (*tovils*) were viewed as a corruption of Buddhism by Hinduism or perceived as pre-Buddhist rituals that had taken on a Buddhist veneer (Scott 1994). Such views ignore that Sinhala offerings of sound and speech to nonhuman beings are conceptualized through Buddhist concepts and values, such as the ranking of beings according to their karma.[25]

When some Europeans emerged as supporters of Buddhism in the late nineteenth century, they tended to praise the religion's supposedly silent nature as an antidote to the unruly noisiness of the modern world. Consider the statement of Helena Blavatsky (1884: 291), cofounder of the Theosophical Society, that, "when resting in Nirvana, the final bliss, Buddha is the silent

*Jim Sykes*

monad, dwelling in darkness and silence." The upshot is that the association of Sinhala rituals with demonism rather than Buddhism marked their *sounds* as "not Buddhist," while "proper Buddhism" came to be defined by the solitary monk in silent meditation—a global icon that, fairly or not, remains to this day. This was a rationalization of Buddhism that placed it outside sonic efficacy, astrology, and drumming. And while doctrinal Buddhism to a large extent *is* located outside these elements, I have shown in this chapter that these phenomena are intimately tied to Buddhism as it is practiced in Sri Lanka.

## CONCLUSION: SOUND STUDIES BEYOND THE LITANY

A part of the colonizing/missionary project involved the splitting of sound from the other senses, with native sounds on the side of demonism/animality and writing on the side of Christianity/culture/ethics. If we are not careful, we will wind up reinstituting this discourse when we apply the divisions of the audiovisual litany (including its definition of the self and public versus private space) through our methodology. John Holt (2004: 350) has noted that rationalized Buddhism has some influence today on urban Sinhala Buddhist nationalist monks from Colombo, who may discredit villagers' appeals to deities (such as Vishnu) for protection; because rural Sinhala villagers were frequently enlisted to fight a war fought in large part for interests articulated in Colombo, such monks delegitimize a key way that villagers sought to protect their sons and daughters during the war. Let us not unwittingly perform similar injustices by assuming that sound can be easily isolated anywhere. Instead, I suggest we consider how different cultures have their own internal *debates* about the relations between sound, speech, hearing, listening, vision, writing, representation, objects, space, astrology, and so on. When we do so, sonic efficacy and exchange will emerge as persisting in some places alongside and even *within* practices that may seem at odds with them. In attuning to their persistence, it will be particularly important to recognize non-Christian ways of defining the self, human-nonhuman relations, and how myriad beings produce ethics and sonic efficacies through listening to and using sound.

### Notes

1. Details of the concert can be found in Jess Harvell, "77 Boadrum," *Pitchfork*, July 10, 2007, http://pitchfork.com/features/article/6645-77boadrum.

2. This interview is published in Hisham Akira Bharoocha, "Boredoms," *BOMB*, July 1, 2008, https://bombmagazine.org/articles/boredoms.

3. The Sinhalas are Sri Lanka's ethnic majority, making up roughly 74 percent of the island's population. The vast majority follow Theravada Buddhism, the kind of Buddhism that is also prevalent in mainland Southeast Asia.

4. An English language version of this material can be found in M. M. Rosana Wasantha, "Horoscope and Astrological Explanation of the 77 Boadrum Performance," *Kendara Sinhalen* blog, August 14, 2009, https://kendaralk.wordpress.com/2009/08.

5. In Vedic astrology, a Hora is a duration ruled by a particular planet.

6. This is correct, since the event occurred toward the end of the much maligned administration of President George W. Bush.

7. As Wasantha's paper shows, Sinhala Buddhist conceptions of sonic efficacy can be mapped onto events in the West.

8. For starters, I have already argued that in Sinhala Buddhist religious contexts, sound is *separate* from the body, although it certainly acts *on* the body. In what follows, I am more concerned with the persistence of sonic protection in the Theravada Buddhist world. Similar examples could be given for Thailand, for example, where Buddhist chant generates protective power in amulets, which then protect the wearer.

9. My argument here thus goes against much canonic writing on the arts in global modernity (see, e.g., Attali [1977] 1985; Rancière 2010) that assumes the arts became globally disenchanted by about the eighteenth century and is more in line with Schmidt (2000). Following the 77 Boadrum example, I suggest that if we more fully locate sonic efficacy in the West's own sonic archive, we will learn that the ocular-centrism and disenchantment of sound in modernity was not as totalizing as is typically imagined.

10. As Heike Behrend, Anja Dreschke, and Martin Zillinger (2014: 15) put it, while efficacious religious practices such as spirit mediumship have "been imagined as a sign of pastness and as a representation of tradition, modernity in fact also produced a new fusion of media and magic." Inventions such as photography, video, sound recording, and digital technology did not rationalize or eliminate efficacious rituals but were incorporated into them, sometimes taking on magical properties themselves. As Rosalind Morris (2000) puts it, spirit mediums and new media share a "technology of the uncanny" that is reproduced anew with the incorporation of new media (Behrend et al. 2014: 15).

11. Sterne's audiovisual litany is a founding document of sound studies, and its tenets are well-known. However, it will be useful to recall a few of its dictums: hearing is spherical, vision is directional; hearing immerses its subject, vision offers a perspective; sounds come to us, but vision travels to its object; hearing is concerned with interiors, vision is concerned with surfaces; hearing is about affect; vision is about intellect; hearing is a primarily temporal sense, vision is a primarily spatial sense (adapted from Sterne 2011: 212). As Sterne (2011: 212) notes, this list is "rhetorically powerful" but not accurate in that it does "not actually hold up when we closely examine auditory experience."

12. Sterne (2012: 13) argues that Ong's aim was "to inject some mystery and transcendence back into a world of thought he considered too concerned with order and reason"; he aimed to do this by better understanding "the conditions under which it was possible for people to hear the word of God" in the modern age. The concept of orality, Sterne (2012: 13) suggests, "has its roots in a spiritualist theological orientation."

13. The exception is Murray Schafer's ([1977] 1993) theorization of "the soundscape" and some work in the field of ecomusicology, which recursively privileges nature sounds in a way that could be understood as a "secular enchantment" (Engelhardt 2014) of sound and the natural world, a retread of Ong's theology without the theology. (For a similar claim, see Ochoa Gautier 2016.)

14. It is from this vantage point that we can grasp how modern definitions of the word "music" (and its equivalent in other languages) played a role in secularizing some "religious" sounds in modernity: defining something as "music" became a way to associate it with the outward expression of the emotional states of an ethnic and religious community, and thus as a practice that emerges into public through performance. In an underappreciated paper, Janaki Bakhle (2008) argues that the classicization of music in northern India was a secularizing process that, in turn, generated the possibility that Hindustani classical music could be "re-enchanted" within the domain of secular, public performances.

15. To be clear, this is not because I imagine the person to be listening to "secular" music, but because the process that generated the moral certitude surrounding this mode of listening (i.e., the celebration of the interiorization of sonic expression in public space) is a formation of the secular.

16. For Buddhists, one's status in life is determined by one's karma, and one's karma (Pali, *kamma*) is determined by the merit (Pali, *puñña*) one has accrued. One can receive merit through gestures of generosity (such as giving food to monks) but also by, say, listening to the Buddha's teachings recited by monks, an act that (as I explain later) has much efficacy.

17. "Blake in a Diplomatic Coup?" *Sunday Island*, online ed., n.d., accessed on July 17, 2018, http://www.island.lk/2008/01/13/politics3.html

18. "Blake in a Diplomatic Coup?"

19. This also includes certain taboo topics such as menstruation.

20. Undoubtedly, a stigma exists against yak tovils today, which missionaries branded "exorcisms" and some view as anathema to Buddhism. But during my fieldwork, I did accompany my low country drum teacher (*gurunnānsē*) to several yak tovils. The stigma seems to have driven the rituals indoors, where they are still routinely performed, just without music and dance (De Silva 2000).

21. For a detailed study of the drumming in the Bera Poya Hevisi, see Sykes (forthcoming).

22. Elizabeth Harris (2006) suggests that British colonial writings on Buddhism can be split into three periods. From 1796 to about 1830, the British primarily considered Buddhism in relation to Brahmanism (Hinduism) because of their prior engagement with

Hinduism in India. Writings in this period included those in the *Journal of the Asiatic Society of Bengal*, founded by the philologist Sir William Jones (1746–94); ruminations by priests such as the Scottish Anglican James Cordiner (1775–1836) and missionaries (beginning with five sent from the Calvinist London Missionary society in 1805); journals of colonial officials, such as the doctor John Davy (1790–1868); and travelogues. The second phase (1830–70) consisted mainly of travelogues, missionary reports, and diaries that largely continued the beliefs of the first period. After 1870 to the end of the century, a more sympathetic and nuanced understanding of Buddhism emerged through the likes of Rhys Davids (founder of the Pali Text Society) and European Buddhist converts, including the exoticist writings of the Theosophical Society.

23. The Methodist missionary William Harvard (1790–1857) stated that nirvana "to the Singhalese in general, conveys no other idea than that of *annihilation*" and "Budhuism, in its original form, is probably the only system of undisguised Atheism ever promulgated; and presents the curious moral anomaly of the founder of a system (who himself denied a Creator) being at length constituted a god by his own disciples. He who rejected all religious worship, as vain and foolish, has now temples reared to his name, in which he is worshipped: and his image is reverenced as a deity, wherever it is seen!" (Harvard 1823: lvi, cited in Harris 2006: 24).

24. The importance of speech to the missionaries is well documented. For example, in the second half of the nineteenth century, a number of well-known public debates were held in Ceylon in which "Sinhala Buddhist orators met Christian missionaries . . . expertly countering charges such as the accusation that Buddhism's atheism breeds moral depravity" (Mahadev 2015: 129).

25. Long ago, Gananath Obeyesekere (1963) noted that both of these systems, doctrinal Buddhism and the "spirit religion," should be seen as compatible and forming a unified system, "Sinhala Buddhism."

### References

Agrama, Hussein Ali. 2012. *Questioning Secularism: Islam, Sovereignty and the Rule of Law in Egypt*. Chicago: University of Chicago Press.

Ambos, Eva. 2011. "The Obsolescence of the Demons? Modernity and Possession in Sri Lanka." In *Health and Religious Rituals in South Asia: Disease, Possession and Healing*, ed. Fabrizio Ferrari, 199–212. New York: Routledge.

Argenti-Pillen, Alex. 2002. *Masking Terror: How Women Contain Violence in Southern Sri Lanka*. Philadelphia: University of Pennsylvania Press.

Asad, Talal. 1993. *Genealogies of Religion: Discipline and Power in Christianity and Islam*. Baltimore, MD: Johns Hopkins University Press.

Asad, Talal. 2003. *Formations of the Secular: Christianity, Islam, Modernity*. Stanford, CA: Stanford University Press.

Attali, Jacques. [1977] 1985. *Noise: The Political Economy of Music*. Minneapolis: University of Minnesota Press.

Bakhle, Janaki. 2008. "Music as the Sound of the Secular." *Comparative Studies in Society and History* 50, no. 1: 256–84.

Bartholomeusz, Tessa J. 2002. *In Defense of Dharma: Just-War Ideology in Sri Lanka*. New York: Routledge.

Bass, Daniel. 2013. *Everyday Ethnicity in Sri Lanka: Up-Country Tamil Identity Politics*. New York: Routledge.

Bate, Bernard. 2005. "Arumuga Navalar, Saivite Sermons, and the Delimitation of Religion, c. 1850." *The Indian Economic and Social History Review* 42, no. 4: 467–82.

Behrend, Heike, Anja Dreschke, and Martin Zillinger, eds. 2014. *Trance Mediums and New Media: Spirit Possession in the Age of Technical Reproduction*. New York: Fordham University Press.

Birla, Ritu. 2009. *Stages of Capital: Law, Culture and Market Governance in Late Colonial India*. Durham, NC: Duke University Press.

Blavatsky, Helena. 1884. *Isis Unveiled: A Master-Key to the Mysteries of Ancient and Modern Science and Theology*. New York: J. W. Bouton.

Canell, Fanella. 2010. "The Anthropology of Secularism." *Annual Review of Anthropology* 39: 85–100.

Chatterjee, Partha. 1986. *Nationalist Thought and the Colonial World: A Derivative Discourse?* Minneapolis: University of Minnesota Press.

Collins, Steven. [1982] 1990. *Selfless Persons*. Cambridge: Cambridge University Press.

Connolly, William. 1999. *Why I Am Not a Secularist*. Minneapolis: University of Minneapolis Press.

Corbin, Alain. 1998. *Village Bells*. New York: Columbia University Press.

Deegalle, Mahinda. 2006. *Popularizing Buddhism: Preaching as Performance in Sri Lanka*. Albany: State University of New York Press.

De Silva, Premakumara. 2000. *Globalization and the Transformation of Planetary Rituals in Sri Lanka*. Colombo, Sri Lanka: International Centre for Ethnic Studies.

Eisenlohr, Patrick. 2011. "The Anthropology of Media and the Question of Ethnic and Religious Pluralism." *Social Anthropology* 19, no. 1: 40–55.

Engelhardt, Jeffers. 2014. *Singing the Right Way: Orthodox Christians and Secular Enchantment in Estonia*. New York: Oxford University Press.

Erlmann, Veit. 2010. *Reason and Resonance: A History of Modern Aurality*. Brooklyn, NY: Zone.

Gamage, Siri. 2007. "Concept of Unitary State and Its Roots in the Sinhala Consciousness: Historical Hints from Michael Roberts's 2003 Book." *Asian Tribune*, February 15.

Gombrich, Richard. 2006. *Theravada Buddhism: A Social History from Ancient Benares to Modern Colombo*. London: Routledge.

Harris, Elizabeth. 2006. *Theravada Buddhism and the British Encounter: Religious, Missionary and Colonial Experience in Nineteenth-Century Sri Lanka*. New York: Routledge.

Harvard, William Martin. 1832. *A Narrative of the Establishment and Progress of the Missions to Ceylon and India.* London: Printed for the author.

Hirschkind, Charles. 2006. *The Ethical Soundscape: Cassette Sermons and Islamic Counterpublics.* New York: Columbia University Press.

Holt, John Clifford. 2004. *The Buddhist Vishnu: Religious Transformation, Politics, and Culture.* New York: Columbia University Press.

Humphreys, Christmas. 2012. *Exploring Buddhism.* New York: Routledge.

Kane, Brian. 2014. *Sound Unseen: Acousmatic Sound in Theory and Practice.* New York: Oxford University Press.

Kapferer, Bruce. 1997. *A Feast for the Sorcerer: Practices of Consciousness and Power.* Chicago: University of Chicago Press

Langdon, Samuel. 1886. *My Mission Garden.* London: T. Woolmer.

Mahadev, Neena. 2015. "The Maverick Dialogics of Religious Rivalry in Sri Lanka: Inspiration and Contestation in a New Messianic Buddhist Movement." *Journal of the Royal Anthropological Institute* 22, no. 1: 127–47.

Mahmood, Saba. 2015. *Religious Difference in a Secular Age: A Minority Report.* Princeton, NJ: Princeton University Press.

Morris, Rosalind. 2000. *In the Place of Origins: Modernity and Its Mediums in Northern Thailand.* Durham, NC: Duke University Press.

Obeyesekere, Gananath. 1963. "The Great Tradition and the Little from the Perspective of Sinhala Buddhism." *Journal of Asian Studies* 29, no. 2: 139–53.

Ochoa Gautier, Ana María. 2014. *Aurality: Listening and Knowledge in Nineteenth-Century Colombia.* Durham, NC: Duke University Press.

Ochoa Gautier, Ana María. 2016. "Acoustic Multinaturalism, the Value of Nature, and the Nature of Music in Ecomusicology." *boundary 2* 43, no. 1: 107–41.

Rancière, Jacques. 2010. *Dissensus: On Politics and Aesthetics.* London: A&C Black.

Reed, Susan. 2010. *Dance and the Nation: Performance, Ritual, and Politics in Sri Lanka.* Madison: University of Wisconsin Press.

Roberts, Michael. 1995. *Exploring Confrontation: Sri Lanka: Politics, Culture, and History.* Chur, Switzerland: Harwood Academic.

Roberts, Michael. 2004. *Sinhala Consciousness in the Kandyan Period 1590s to 1815.* Colombo, Sri Lanka: Vijitha Yapa Publications.

Sartori, Andrew. 2008. *Bengal in Global Concept History: Culturalism in the Age of Capital.* Chicago: University of Chicago Press.

Schafer, F. Murray. [1977] 1993. *The Soundscape: Our Environment and the Tuning of the World.* New York: Simon and Schuster.

Schmidt, Eric Leigh. 2000. *Hearing Things: Religion, Illusion, and the American Enlightenment.* Cambridge, MA: Harvard University Press.

Scott, David. 1994. *Formations of Ritual: Colonial and Anthropological Discourses on the Sinhala Yak Tovil.* Minneapolis: University of Minnesota Press.

Seneviratne, H. L. 1999. *The Work of Kings: The New Buddhism in Sri Lanka.* Chicago: University of Chicago Press.

*Jim Sykes*

Sterne, Jonathan. 2003. *The Audible Past: Cultural Origins of Sound Reproduction*. Durham, NC: Duke University Press.

Sterne, Jonathan. 2011. "The Theology of Sound: A Critique of Orality." *Canadian Journal of Communication* 36, no. 2 (summer 2011): 207–25.

Sterne, Jonathan, ed. 2012. *The Sound Studies Reader*. New York: Routledge.

Stewart, James. 2017. "Dharma Dogs: Can Animals Understand the Dharma? Textual and Ethnographic Considerations." *Journal of Buddhist Ethics* 24: 39–62.

Sykes, Jim. 2015. "Sound Studies, Religion, and Public Space: Tamil Music and the Ethical Life in Singapore." *Ethnomusicology Forum* 24, no. 3: 380–413.

Sykes, Jim. 2017. "Sound as Promise and Threat: Drumming, Collective Violence, and Colonial Law in British Ceylon." In *Cultural Histories of Noise, Sound and Listening in Europe, 1300–1918*, ed. Ian Biddle and Kirstin Gibson, 127–51. New York: Routledge.

Sykes, Jim. 2018. *The Musical Gift: Sonic Generosity in Post-War Sri Lanka*. New York: Oxford University Press.

Sykes, Jim. Forthcoming. "On the Sonic Materialization of Buddhist History: Drum Speech in Southern Sri Lanka." *Analytical Approaches to World Music*.

Taylor, Charles. 2007. *A Secular Age*. Cambridge, MA: Harvard University Press.

Weiner, Isaac. 2013. *Religion Out Loud: Religious Sound, Public Space, and American Pluralism*. New York: New York University Press.

White, Shane, and Graham White. 2006. *The Sounds of Slavery: Discovering African American History through Songs, Sermons, and Speech*. Boston, MA: Beacon.

# 10

# "Faking It"

## MOANS AND GROANS OF LOVING AND LIVING

## IN GOVINDPURI SLUMS

*Tripta Chandola*

.....

## INTRODUCTION: ON LISTENING TO THE
## LISTENED OF THE LISTENER

I begin with a quote: "Quite simply: listen to it" (Feld 2012: xxvii).[1] This rather commonsensical assertion by Steven Feld to navigate through his work with the people of Bosavi is pregnant with possibilities for navigating the acoustic ecologies we "know and [make a] habit of knowing" (Feld 2012: xxvii). At this stage in my research trajectory, I am often introduced as an ethnographer who specializes in listening. This framing always tickles me, for how does one specialize in listening when—unless compelled by considerations of paracusis of varying orders—listening as a function of a hearing person is an involuntary act? Because for such people, the ears do not close. Thus, when I make a call for listening, with its methodological and political deliberations, is there not implicitly an assumption of intentional deafness in these regards—a closing of the ears, so to speak?

These and other, similar conundrums highlight the challenges in actualizing the task of listening. This chapter is an attempt to respond to key questions raised by the editors of this volume, such as: "What do the entanglements of sound in various regimes of value, such as practices of listening in empire and regulations of sound in colonial and postcolonial contexts, tell us about the history of sound in modernity? And how can studies of the structural and sensory marginalization of the urban poor under contemporary neoliberal conditions contribute to our understandings of what sound is and what it does?"[2] There is nothing simple about this.

The ethnographic research experiences in the slums (*jhuggis*) of Govindpuri on which I draw here span more than a decade and bear testimony to the challenges of this undertaking.[3] The fulcrum to engage with these experiences is the practice of listening and privileging what, how, when, and why the slum dwellers themselves listen in and into. The matter of agreeing on listening (and listenings) as a trope of engagement with the slum dwellers about their everyday lives, though, was neither intentional nor strategic. My first encounter with the dense corporeality of the everyday in the slums — its clogged drains, overwhelming smells, crowded lanes, and "obscene" loudness — came as a young researcher in early 2004 for a project funded by the Department for International Development. By then I had lived in the city of Delhi for almost a decade, attended a left-leaning university, and worked on projects that focused on marginalized sections of society in the city. In short, I was well versed and engaged in the academic and activist grammar about the violence and disenfranchisement of the slums and their residents, but I had never encountered their reality. I had never paid a visit to any slums in the city.

Thus, as a researcher and interlocutor, I found myself initially quite unsure about how to listen in and into the slums. Lacking the vocabulary to engage with people's everyday lives demanded humility of me to learn to listen, and this listening revealed the highly evolved sense of "self" of the slum dwellers in regard to their identities, the spaces they inhabit, and their engagements with their identified Others — both within the slums and in the broader ecology of the city. This chapter emphasizes listening(s) of the slum dwellers into their everyday practices, as articulated and complicated by them, to present insights into the wider social, cultural, spatial, emotional, sensorial, and political cosmos of the slum, often left unheard and unacknowledged in both mainstream and academic discourse. It is within these coordinates (spatial, social, cultural, and political), I suggest, that the diverse instances of listening (collective and individual) I discuss are lodged.

## IN WHICH SONGS BREAK THE AFTERNOONS

In 2006, I found myself hanging around with different groups in Govindpuri's camps, depending on their availability, generosity, and mutual curiosity. The protagonists of the incident I recount here are a group of young, vivacious girls ages sixteen to nineteen. I was only a few years older than them, but in my researcher role as a middle-class, educated person I was given credit, however misplaced, for having a higher, worldly wisdom; and since then I have been universally known as *didi* (elder sister) in the camps by young and old alike. The girls were restrained at first when they told me the following story, but they realized soon enough that most of my so-called wisdom was not moralistic or judgmental. In fact, for them and for me, the most engaging and entertaining times came when we discussed the "boys": the loves; the professed ones; the stalkers; the charming, uncouth ones; those worthy of being considered husband material; and of course, the liars and the cheats.

Most of the girls, while giving the impression of being *chaalu* (street smart) in their dealings with boys, betrayed their ages, lack of experience, and insecurity when our conversations steered toward the "liars and cheats." However, during one such discussion, Priyanka, a young girl with a wicked sense of humor, remarked, "In fact, it is not the boys who are to be dreaded and feared for lying and cheating but one among our own." She then pointed to one of the girls, a close confidant, who, she said, had to be watched because she could turn into *aasthin ka saap*—a serpent in the bosom, capable of vicious betrayals and lies. Hearty laughter followed, and the girls subsequently recounted the following incident at great length, with zestful interjections and annoyance feigned by the target of the so-called betrayal, Priyanka herself.

And the story went this way: a couple of years earlier, Priyanka, who was then about seventeen, was besotted with a boy of the same age who lived a few lanes away in the Navjeevan Camp. She revealed her attraction to the girls, her confidants, whom we can identify as accomplices, *aasthi ki saap*. For the next few months, the girls found themselves completely engrossed by this affair, however one-sided.

One of the girls' brothers owned a tape deck complete with amplifier and speaker boxes. The girl was allowed to use it when her brother was not around (or she did so discreetly without his knowledge). The group of girls would congregate in her house, usually in the afternoon, when the men were away and the women were preoccupied elsewhere, to listen to their favorite songs. Priyanka, enamored as she was with the boy, would insist on playing romantic songs while regaling her friends with talk about her prospective lover, speculating about what he was like and what caught his fancy, and pestering the others to tell her that he was equally enchanted with her.

At this juncture, a digression is necessary to contextualize the gendered and technological ecologies in the slums of Govindpuri in 2006. The techno-

*Tripta Chandola*

logical landscape at that time was marked by limited penetration of mobile phones, which were still considered a luxury because of both the high cost of the handsets and the difficulty in securing a connection, which required the presentation of several identification documents. The slums of Govindpuri remain a highly gendered space where the social, spatial, and technological mobility of women—especially young, unmarried ones—is carefully monitored. In the past decade, the spaces and possibilities, and thus the mobility available to the women, has witnessed a significant shift. In 2006, therefore, the group of girls in the story did not enjoy the freedom that is available today. Most of them were school dropouts and were not permitted to venture outside the slums unescorted—neither for pleasure nor to seek employment—and a romantic alliance made public would invite the wrath of family and community.

As the days wore on, the girls became tired of Priyanka's performance, especially since all of them knew it would go nowhere—or, more accurately, that it would not be allowed to go anywhere. This is when the girl with the tape deck proposed an ingenious scheme, which in the end won her the title "biggest and vilest of the serpents in the bosom," and why, at the time the story was recounted, she was the most *chaalu* of the girls. The scheme was rather simple, exploiting the fact that the facts could never be confirmed by Priyanka or the object of her affection.

Over the next couple weeks, the girls convinced Priyanka that the girl with the tape deck's brother was well acquainted with the boy in question and could be convinced to gently prod the boy about his feelings for Priyanka. Of course, Priyanka was elated; her enthusiasm was not deflated even when the girls informed her that this would come at a price: fifty Indian rupees ($0.73 USD at today's exchange rates). They were vehement in their insistence that the money was not for their benefit but, instead, was to be given to the girl with the tape deck's brother, a classic example of a selfish, manipulative man. Priyanka could not help but imitate the group of girls while recalling this story, demonstrating how well they had performed the example of men being liars and cheats. Collectively they would sigh, Priyanka said, letting out an affected, long-drawn-out moan as part of her imitation—"Oh, this patriarchal world, what choice do we helpless women have?"—to carefully concede that there was "no way that the boy was anything like that."

The girls were content with their scheme, but it only encouraged Priyanka, and she was unrelenting in asking what the boy had said about her and how he said it. As the weeks passed, she became more demanding, urging immediate and definite responses. "Would he take to passing along her lane more often?" she asked. "What color did he think suited her best?" Priyanka had tried a new hairdo and was sure the boy had glanced at her. "What did he think?" she wanted to know.

Eventually, the afternoon indulgence of listening to songs began to lose its frivolity and assumed a gravity that was unsettling for the girls. When she did not receive any response from the boy, Priyanka began to grieve and insisted on playing sad songs about love and its loss. It was then that the girl with the tape deck pulled a trump card from the deck: she proposed yet another scheme both to placate Priyanka and make the girls more money. "Of course, we deserved the extra money," one of the girls told me. "After all, we were the ones who had to endure afternoons into evenings of Priyanka's moaning and groaning."

After waiting a week or so, Priyanka was summoned earlier than usual to the girl with the tape deck's house for their afternoon congregation for songs and confessions. There she found the girls in an excited state, with a peppy dance number enlivening their moods. As she entered, the song that was playing was stopped, and a cassette, which Priyanka did not recall having seen before, was taken out and dramatically inserted into the player. An old Bollywood song of love and yearning resounded in the room. "*Tum ko dekha to yeh khayal aaya zindagi dhoop to tum ghana saya,*" went the lyrics, which can be loosely translated as, "Since I saw you, I realized you are the shaded corner of comfort to my life, which is otherwise burning under the scorching sun." Priyanka—slightly bemused but keenly interested—was informed, with dramatic sighs, pauses, and equivocations, that this was the message the boy in question, the object of her affection, had sent her. Of course, Priyanka's joy knew no bounds. "Could she send him a message back?" she asked.

The celebratory mood in that one-room tenement in the camps of Govindpuri on a mild afternoon suddenly stood as still as an unwanted, uncertain visitor. The girls, by now skilled at their art, orchestrated a collective sigh as a precursor to the announcement that, yet again, a payment would be required. They again evoked the girl with the tape deck's brother as the exemplification of the "vileness" of which men are capable, only to add that the boy Priyanka liked was beyond such behavior. The brother, the group of girls said, had agreed to act as a courier between the two lovers, and he wanted a remuneration from both sides. Priyanka was offended at the thought that the object of her affection who was above such manipulative tactics, would have to pay to profess his love to her. She made some quick calculations and boldly announced that she would pay for the courier on both sides—ten ru-

*Tripta Chandola*

pees per trip. The celebratory mood returned, especially among the group of girls, who had not anticipated such an amazing turn of luck.

Thus began the "fake" real love affair articulated and realized only through songs and listening. The afternoons had a new meaning for Priyanka and the group of girls. It was they, after all, who were acting as facilitators of the affair, at once meticulously choosing songs as messages from the boy and deliberating with Priyanka to choose one in response. They included, but were not limited to, songs of yearning, angst, penetrating questioning, jealousy, and reproach.

This state of affairs continued for almost a year, but eventually it faded. While recalling the incident for this chapter, I could not remember whether I had asked Priyanka, or the rest of the girls, who were cackling without showing that they recognized their deceit, why they had done it. What I do distinctly remember, on the verge of turning forty, is having felt privileged for being invited to share, to *listen into*, the revelry of those afternoons—the humming of the songs, the deliberations, the delight, the pangs of unfulfilled desire, the innocence—and the mutual realization, by the girls and myself, that I was, after all, not so wise or any different from them in things that mattered.

I am still in touch with one of the girls, Pooja, who is now a grown woman of twenty-nine and a mother of two young children. I recently called to ask her some of the previously unasked pertinent questions, such as, "Why didn't Priyanka question the entire modus operandi? Why didn't she insist on other means of communication, such as letters or a secret rendezvous, especially since a channel had opened and there was a courier for whose services she was paying? Why did the girls continue with the deception, as the money they received from the scheme was not making them rich, and, moreover, they were spending it on collective indulgences?"

Pooja's hearty laughter, resounding of the innocence of those afternoons and recalling that innocence to me years later, filled the room. She said:

> I have not thought about those days in such a long time. In fact, I'm not even in touch with most of the girls. We all got married, one after another. But I don't think you really understand what those afternoons meant for us. Do you? Perhaps neither did we. We all knew, instinctively, deep down, what lay in our fate. We all knew, even Priyanka, that all of it was a big, fat lie, a well-executed charade. Why did Priyanka not insist? Why did we continue? Because listening to those songs, talking about love that was not there and that we knew

would never be there, allowed us all to feel the love, to be and behave like we never could otherwise. I think the magic of it started fading away when one of us, I think it was Seethu, got married—of course, arranged—and the reality started to sink it. We were next in line, and the futility, all the role-playing . . . did not make sense anymore. But that "fake" love and loving was the first for all of us, and perhaps for some of us, it was the only love. That is why we did not hold anything against anyone.

As I ended the call, after exchanging courtesies about the welfare of the children and promises of visits in the near future, I was left wondering about the neutrality of soundscapes. I was left wondering about the social, political, emotional, and moral possibilities available across structures of gender, community, and caste—especially among the disenfranchised and marginalized in society.

If the ears do not close, and cannot be closed, does that leave listeners entirely stripped of their agency? Were those listening into the afternoons, so nostalgically reminisced about by one in the group of girls, to the songs of love and desire, and feeling love, indeed "fake"? In performing the dualistic, displaced *listening*—of both lover and loved—the girls could claim romantic experiences that otherwise were strictly prohibited and thus unavailable and in doing so, in their own ingenious, cacophonic listening, however momentarily, rupture the hegemonic, patriarchal, and restrictive normative sphere.

## IN WHICH THE EARS ARE INTERCHANGEABLY TUNED

Beedi is one of my closest confidants, friends, and research associates. I met her a few years ago, serendipitously on a hot afternoon in July. Since then, both of us, at different times, have been each other's anchors, providing needed intellectual stimulus and emotional support. Beedi is a feisty, uncompromising (some would say difficult), and enterprising woman. She is also a *jhuggi-waali*, a resident of the slums. Since I first met her in 2008, she has moved out of the slum settlement to Govindpuri, the neighboring lower-middle-class area (the adjoining slums are referred to as the slums of Govindpuri). The tale of that shift—spectacular, intriguing, and revelatory of the issues of mobility across constructions of gender, class, and spatiality with which the slum-dwellers have to deal—demands its own telling, and the scope of the present chapter does not permit it the attention it deserves.

*Tripta Chandola*

However, significant to this particular narrative is the identification as a jhuggi-waali that Beedi continues to claim and assert.[4]

Here I want to invite readers to accompany me on a walk with Beedi through the market (*haat*), which occupies a large space somewhat bridging the gaps between the slum settlement and the lower- and middle-class neighboring areas.[5] The haat is held every Wednesday and is thus referred to as the Budh Bazaar (Wednesday Market). It is open from early evening into late night and spreads out along two kilometers of the main arterial road that passes through Govindpuri, the legal settlement, and runs along the slum settlements, intersecting with a few middle-class apartment blocks and an adjoining settlement, leading toward one of Delhi's (possibly India's) largest unauthorized colonies, Sangam Vihar. The traffic along this road on most days is dense, but when Budh Bazaar is scheduled, it assumes a choreographed opacity that is surprisingly fluid to navigate. Both sides of the roads are resplendent with wares of varying designs—utilitarian and tempting. The goods are arranged on carts, ferried on bicycle paniers, laid on the ground along the road, spread out on plastic sheets, or sold by salesmen on foot.

It was amid these spectacular distractions that I undertook the walk with Beedi. Lacking a definite agenda, we strolled around the markets, giving in to our temptations, fondling items that we did not need, and partaking of street food. Weak attempts to bargain and clever, self-congratulatory spotting of items, especially discounted and secondhand items, were being negotiated sonically through multiple conversations, sellers at once selling and advertising wares, with parallel attempts happening at close quarters.

In between all of this Beedi was regaling me about an online "lover" with whom she had established contact via Whatsapp. He lived in Dubai, and she had never met him, but they spoke on the phone at regular intervals. After a couple months of "friendly" banter, the conversations moved toward exchanges of a more intimate nature. The man on phone, as Beedi referred to him, was an abstracted, distant, and detached fantasy. Even though the voice extended promises of everlasting love, marriage, and togetherness, Beedi was clear that all she wanted from him was the latest iPhone. As a woman of the streets in an earlier life, Beedi was equally aware of what man on phone desired: *phone pe sex* (phone sex).

The contact thus set in motion a cycle of articulated desires and attempts to actualize them on an abstract, mediated plane where the only point of connect was the voice. Beedi acknowledged that the iPhone was never going to materialize, saying, "*Kabhi nahin hota, par socha riks le lete hain, aur hum log bore bhi hote the, tho maja aata tha uski lene mein*" (It will never happen, but where's the harm in taking the risk? And anyway, we used to get bored, so we had fun collectively mocking him). The routine that followed went this way: Beedi and a few of her friends would answer the man on the phone's call and put him on the speakerphone; after a few customary exchanges, he would initiate phone sex, with very exact demands about the kind of sonic performances he desired (deep voice modulations, refrains about desiring him, and, eventually, loud, satisfied moans).

This mediated sonic-sexual encounter was collectively experienced by Beedi and her friends; while it was "fun" to hear him perform phone sex, the desired performance in response quickly became boring for the women. However, man on phone had been indulged, and the promise of the iPhone was still dangling out there, so the girls took turns participating and sonically performing in this abstracted sexual encounter. "After all," Beedi rationalized, "how was he to know who was moaning?"

The girls could be cooking, cleaning the house, or conducting other everyday household chores with the "lover" on the speakerphone articulating and performing his most intimate desire. The women's awareness that man on phone assumed they were intensely involved in the "act" while their response to it was, at best, uninterested and distracted (not to mention shared) is what cracked them up. Here, amid a "cacophonic encounter" of its own merit, the listener (the long-distance lover) is divested of his position in the hierarchy of sonic performances by being denied engaged listening — "a mad man laughing at the rain," as Beedi colloquially summed up man on phone's predicament.

It was amid all of these sonic distractions that technical matters of listening announced themselves on our walk via a singing salesman at one end of the haat, near the Govindpuri metro station. A fairly large crowd was gathered around the stall, and we found ourselves gravitating toward the singing salesman, without knowing what it was he was selling. Undoubtedly, his voice could carry a melody, but he was also singing old Hindi songs of yearning, desire, and loss that, after the sonic boisterousness of the walk along the haat, felt like an orchestrated crescendo and a logical conclusion to this sonically intense almost summer evening.

I was completely taken with the singing salesman. He was wearing a crisply ironed, button-down shirt and trousers; had he not been singing along to a battery-operated *dabha* (music box) placed precariously under the cart, I would have taken for a senior clerk in a desolate office, perhaps sharing Fernando Pessoa's (2002) fate. Beedi, however, was not so taken with the salesman's sonic presence, his performance and punctuation of the haat's

*Tripta Chandola*

soundscape. She announced, not so quietly, to the entranced crowd that she thought he was "faking it"—that it was, in fact, the dabha that was singing, and the salesman was "only moving his lips."

I found myself momentarily unsure about my own listening. A significant premise of my research, after all, is focused on listening into others as listeners, which in this case meant lodging my listening firmly in Beedi's ears. I was at once intrigued and irritated by the sonic dissonance that the singing salesman's performance had created in the "listening to the 'listened' of the listener." Instead of merely contemplating the anxiety caused by this dissonance in the listening and a disruption (of sorts) in the soundscape of the haat—which until this moment I had considered a collective and shared experience at a theoretical, intellectual, and even political level—I decided to put our ears and our listening to the test.

I waded through the crowd, which consisted exclusively of men,[6] and, addressing the singing salesman, put forth Beedi's accusation—a complete denial of his performance. "She believes you are faking it," I said. The transformation in both Beedi's and the singing salesman's persona was remarkable, as their respective sonic positionalities—his as a producer and hers as a listener—were challenged; the salesman shrank a bit and bent down to switch off the music box while insisting that the voice was his own and the box was providing only the background score. Beedi, who is usually feisty, unrelenting, and rarely apologetic, stood staring like a deer in headlights.

I offered a sonic intervention to mitigate the dense silence that had resulted from this dissonant listening in the soundscape of the haat. "Why don't you sing an acoustic version of this lovely lady Beedi's favorite song?" I asked. Beedi feigned embarrassment at being the center of attention but could do little to hide her pleasure at having her listening regarded as important; it had, after all, become the fulcrum to validate the singing salesman's sonic presence and performance. The salesman was more than eager to oblige, if only to reclaim his position (and performative punctuation) in the haat's soundscape. He took a moment to compose himself, then sang a song full of yearning, desire, and misunderstanding. This choice was not arbitrary, as he was attempting to assert (and reclaim) his position in the haat's sonic landscape, in which he had—until just a brief moment ago—occupied a prominent position.

The crowds were enthralled after his performance, demanding more and shouting out requests. Beedi extended profuse, flirtatious apologies for her mis-listening, insisting that it was a ploy to hear him sing for her. I recorded the performance and the interaction. As we started to leave, the salesman asked, rather perfunctorily (realizing, it seemed, that along with singing he also wanted to sell the wares on display), "Won't you take anything from here?" Displaying the recorder to the crowd, I answered, "I am taking your voice with me." A round of applause followed, and in the distance, we could still hear the singing salesman's voice reverberating.

The salesman's name was Md. Tahir, and by day he worked as a clerical assistant at a food-processing factory in the nearby Okhala Industrial Estate.

## CONCLUSION: THE REALITY OF "FAKE" LISTENING IN THE SLUMS OF GOVINDPURI

In this chapter, I have sought to highlight different and diverse instances of "fake" listening by the residents of Govindpuri and, in the process, identify the agency of slum dwellers to claim their listening, soundscapes, identities, and spaces. The instances of listening I have recounted here are not necessarily in contest or conflict with the established, hegemonic orders that organize soundscapes but should be engaged with as listening practices that represent an ethical cohabitation across soundscapes, spaces, communities, and ideas. The extent to which these listenings can be appropriated to exert political, social, and cultural presences and demand structural recalibrations should not be overestimated. Elsewhere, I have insisted that soundscapes are not neutral, apolitical background matter; their recognition as "noisy, silent, musical" are socially, culturally and morally informed and motivated and, thus, essentially political in nature (see Chandola 2012).

The sonic encounters I have discussed in this chapter do not evoke the limitations—social, structural, political—that the residents encounter in their everyday lives. Nor do they celebrate the "heroic entrepreneurship" of the residents to overcome hardships. Instead, in this chapter I have engaged with some typically sidelined aspects of poor, disenfranchised, and marginalized sections of society—their emotional, "frivolous," tempestuous indulgences—with a focus on the ingenious ways they "fake it," as manifested in their listening.

### Notes

1. The heading of this section is a play on Dürrenmatt's (1988) book title, *The Assignment, or, On the Observing of the Observer of the Observers*.

2. This question appears in an early version of the paper that would eventually become the introduction to this volume; see Gavin Steingo and Jim Sykes, "Remapping Sound Studies," *Humanities Futures*, https://humanitiesfutures.org/papers/remapping-sound-studies.

*Tripta Chandola*

3. *Jhuggis* in popular evocations are understood to "represent settlements typically marked by some measure of physical, economic, and social vulnerability . . . these settlements are often called 'slums.' Within planning paradigms in Delhi, however a 'slum' refers specifically to a settlement designated as such under the 1956 Slum Areas Act" (Bhan 2016: 40). This act declares slums areas as settlements which "by reason of dilapidation, overcrowding, faulty arrangement and design of such buildings, narrowness or faulty arrangement of streets, lack of ventilation, light or sanitation facilities, or any combination of these factors, are detrimental to safety, health or morals" (Slums Areas Act 1956). The official definitions of the slums enters the everyday imagination and translations in such a manner that the "structural" factors, particularly the lack of morals, are employed to exert practices of *othering* on the residents of the *jhuggis*, with implications of social, physical, political nature and denial of agency.

I use the terms used by the residents themselves to refer to the settlements, and that is either *jhuggis* or camps. In their usage, *jhuggi*, first and foremost, is the site and space of their settlement, habitation, and belonging. Collectively and popularly referred to as the Govindpuri *jhuggis*, the settlement I am concerned with in this chapter is located in South Delhi and consists of three different "camps": Nehru Camp, Navjeevan Camp, and Bhumiheen Camp. Adjoining the slum settlements is a lower-middle-class area also known as Govindpuri. The three camps fall under the Kalkaji constituency, which is one of the seventy Vidhan Sabha (Legislative Assembly) constituencies of the National Capital Territory.

4. This introduction to Beedi is necessary to establish the spaces of sociality across which the two of us engage and negotiate. While her status and vehement self-identification as a jhuggi-waali and my background as middle class and educated does not affect our "everyday" relationship, neither one of us sustains amnesia about our positionality on the broader structural plane, which has a significant impact on the capital—social, cultural, political—we can accrue on that account. I also need to establish that Beedi is not my sole point of entry to, and engagement in, the slums of Govindpuri, as in the past thirteen years I have established strong social relationships in the three camps. However, it is more often than not Beedi who provides the keen insights and perspective as a cultural translator that allow me a nuanced, contextual understanding of the everyday in the slums.

5. Beedi inquired about the rationale for this undertaking and asked what I was trying to establish in doing so. I drew the outline of the argument, insisting on the manner in which a specific kind of listening into spaces identified as Othered renders them noisy and thus are attempts to silence them. "That might be the case," she responded, "but we are not particularly the silent types or the ones who can be easily silenced, are we now?" This brief response succinctly captures the sonic self-awareness of the spaces and presences that can be claimed and asserted through soundscapes.

6. This is not surprising, given that the wares for sale were for men, and the event recounted took place in the late evening.

## References

Bhan, Gautam. 2016. *In the Public's Interest: Evictions, Citizenship, and Inequality in Contemporary Delhi*. New Delhi: Orient Blackswan.

Chandola, Tripta. 2012. "Listening into Others: Moralising the Soundscapes in Delhi." *International Development Planning Review* 34, no. 4: 391–408.

Dürrenmatt, Friedrich. 1988. *The Assignment, or, On the Observing of the Observer of the Observers*, trans. Joel Agee. Chicago: University of Chicago Press.

Feld, Steven. 2012. *Sound and Sentiment: Birds, Weeping, Poetics, and Song in Kaluli Expression*, 3d ed. Durham, NC: Duke University Press.

Pessoa, Fernando. 2002. *The Book of Disquiet*. London: Penguin.

*Tripta Chandola*

# 11

# Disorienting Sounds

## A SENSORY ETHNOGRAPHY OF

## SYRIAN DANCE MUSIC

*Shayna Silverstein*

.....

One summer afternoon in central Damascus in the late 2000s, a young man and his friend arrived at my flat to teach *dabke*, Syria's national performance tradition, to a friend and me — both of us visiting from the United States. Mohammad and I had met through an emerging *capoeira* scene in Syria's capital, and I had had the opportunity to watch him improvise dabke at a birthday party of a fellow *capoeirista*.[1] He was delighted to share his individual style with us, although he offered a disclaimer that his performance techniques were not quite authentic: his family traces their origins to urban Damascus, and like most Syrians, Mohammad maintained that dabke was a rural, pre-Islamic performance tradition brought to Damascus through rural-urban migration as recently as the 1930s. He acquired his skills by participating in numerous *ḥaflāt* (parties saturated with this popular dance music), at which he and other young men displayed their youthful masculinity through dabke performance. However, the informal lesson flopped. Despite our extensive dance training in the United States (ballet, West African dance, and tap dance, respectively), my friend and I felt clumsy as we awkwardly tried to mimic Mohammad's technique. He also felt frustrated at not being able to break down his technique into digestible learning bites. Realizing that his performance was dependent on the skill of other participants, he excused himself at the end of our session with a profuse apology: "I'm so sorry! The *mazāj* [atmosphere; literally, 'mood'] is just not there. You know, I need someone to hold down the rhythm for me. Then it would work out."

Mohammad's sentiment of failure developed from, I suggest, the lack of atmosphere that he identified at the end of our session. Atmosphere (and the lack thereof) is a performance condition related to the embodied experience of "flow," a process that Mihaly Csikszentmihalyi (2008: 65) conceptualizes as a "field of force" in which the self loses consciousness in the very process of expanding its boundaries. Mohammad was not able to "flow" with us because we did not possess the technique, or embodied knowledge, that would anchor his dabke movements. According to ethnomusicological literature of the Arab world, the phenomenon known as *salṭana* embeds the (admittedly universalist) concept of flow in and as culturally specific mediations. In the art music tradition of *ṭarab* performed primarily in Syria and Egypt, for instance, acts of listening are considered central to how musical aesthetics register affectively in the Arab world. Musicians try to achieve salṭana in the course of performance to arouse emotions among listeners. As Jonathan Shannon (2006) writes in his ethnography of *sammīʿa* (musical connoisseurs) in Aleppo, ṭarab is "an index of the social relations of musical performance" that is mediated by the performativity of listening as much as the performatics of the musicians.[2] In contrast to ṭarab's world of serious listeners in which sonority is paramount, Syrian dance music constitutes a strikingly different "social field of listening" (Kapchan 2017: 1), in which sound, affect, and the participant's body "fold in on one another" (Kheshti 2015: 55).[3] Listening to dabke, among other genres of Syrian dance music, is less about the aesthetic structures of sound than about an intensely somatic environment in which the senses constitute a totality through which the body perceives the world.

In what follows, I examine performance dynamics in Syrian dance music to argue for what the introduction of this volume, borrowing from Jonathan Sterne (2012), calls a "conjunctural sound studies," or the consideration of how nonauditory senses may transform the relationship between the listener and the sound object. Syrian dabke practices are an exemplary case study for taking up a redistribution of the senses within sound studies as they demand consideration of nonauditory senses in ways that help to better understand the performance dynamics of dance music, which are related to but ontologically different from listening-centric practices such as ṭarab. Moreover, this set of issues challenges how sound studies tends to structure the senses hierarchically—namely, to privilege audibility over visuality and to disregard what other senses, including the haptic, tactile, olfactory, and proprioceptive, contribute to our understandings of sound.[4] Given the highly participatory and interactive dynamics of dabke as a social dance

*Shayna Silverstein*

performed to specific genres of Syrian popular music, this essay will focus on kinesthetics to raise questions about how proprioception, movement, and tactility direct bodies in sonically dense environments.

My focus on kinesthetics is also a tactic for engaging "somatic modes of attention" (Csordas 1993) within sound studies, a field that, like much of academe, tends toward somatophobia. As Bryce Peake (2016: 80) points out in his study of listening practices and subject formation in British Gibraltar, sound studies tends to "render listening without bodies . . . and neglect the social relations of force that constitute the varying social positions from which specific listeners listen." When sound studies does attempt to account for positionality, as Tim Ingold (2000) writes, it too often puts racialized and gendered bodies to work by asking them to account for the visual episteme of Western imaginaries. Mapping certain sensibilities onto non-Western Others to dismantle the visual episteme is problematic because it asks Othered bodies to tackle certain divides between non-West and West, and aurality and visuality. Although I acknowledge that the ethnographic fieldwork represented in this chapter can be traced to these histories of colonial encounter, I try to reorient which bodies do the work by taking up Deidre Sklar's proposition to explore embodied knowledge through the ethnographer's sensibilities. In her movement ethnography of a religious fiesta in Tortugas, New Mexico, Sklar (2001a) articulates how the body is not a text that can be read; rather, we can explore physical knowing through movement through the phenomenological body, of the researcher as well as the interlocutor. In what follows, I stage scenarios of ethnographic encounter that present fieldwork as a "dialogic performance" in which, as an ethnographer, I become a "co-performer" with my interlocutors, equally invested in the "range of yearnings and desires" (Madison 2011: 186) that constitute subject formation.

Fieldwork-based encounters are also critical for disrupting and disorienting the ethnographer's sensory modes of perception. As Brian Larkin (2014: 996) points out, perception is not an "invariant cognitive process"; rather, the contingencies of lived experience entrain our bodies to perceive the world in culturally specific ways. In *Queer Phenomenology*, Sara Ahmed offers an example of how perception affects the moving body. She argues that the surfaces of our bodies acquire their shape not because "bodies simply have a direction, or that they follow directions. Rather, in moving this way rather than that, and moving in this way again and again . . . bodies are 'directed' and they take the shape of this direction" (Ahmed 2006: 15–16; cf. Albright 2013: 14). We can disrupt the "habitual and elided" relations through which "we end up seeing simply what is in front of us" (Ahmed 2006: 1; cf. Albright 2013: 13) through the work of sensory disorientation.

Through what I call "sensory disorientation," ethnographic immersion in the field "slants" (Ahmed 2006) these reiterative processes so they become more open-ended rather than habituated or rote. As a method of performance ethnography, sensory disorientation seeks to both engage in and disrupt sociocultural practices to better understand the cultural logics and performative processes that shape ethnographic subjectivity. For instance, the opening encounter among Mohammad, our friends, and me is an example of how sensory disorientation functions as method. By disorienting myself and my friend, and defamiliarizing dabke for Mohammad, we were able to "simultaneously recognize, substantiate, and (re)create ourselves as well as Others through performance" (Madison 2011: 166). The effects of performance ethnography are to better understand how our methods imbricate our own bodies. Disorientation of the ethnographer's sensibilities informs how we approach sound and the insights we gain through its study because it accounts for body techniques, such as listening and dancing, as incidental, subject to spatial-temporal disjunctures, and indicative of social distinctions between researchers and their object of study.

In what follows, I provide a brief background on the performance tradition of dabke and the social events at which it is practiced, including the sonic mediations that typically take place at these events. I then stage several encounters between myself and interlocutors that pivot on moments of sensory disorientation to convey key performance dynamics of dabke. That these encounters took place in prewar Syria situates this essay in a particular historical juncture prior to the mass migration of Syrians seeking refuge from civil war. By embodying the ethnographic process through dabke practice, I aim to deprivilege intellective modes of knowledge production and redistribute the senses in ways that challenge the disciplinary genealogy of sound studies.

## ON DABKE

The performance tradition of dabke is widely practiced across the region known as al-Mashriq, or present-day Syria, Lebanon, Jordan, and Palestine. A curvilinear group dance performatively linked with an upbeat dance music genre of the same name, dabke is typically practiced at wedding celebrations, restaurants and nightclubs, and other social settings. It is considered a rural tradition that was urbanized in the course of mass migration into Syria's

*Shayna Silverstein*

cities in the 1950s and 1960s. During the rise of Arab nationalist movements in the mid-twentieth century, modernizing elites popularized and politicized dabke, transforming it into a folk dance that embodied the nation-state. In Syria, the first state-sponsored folk dance troupe emerged out of the political culture of the short-lived United Arab Republic, a Syrian-Egyptian political alliance (1958–61). Syrian youth traveled to Cairo to perform their rural traditions for Pan-Arab nationalistic audiences at political events during this period. While staging claims for Arab unity, these performances also reinforced the logic of the Syrian homeland, which sought to consolidate vast regional and ethnic differences through the performativity of the pre-Islamic and premodern Arab subject.[5]

Performance events typically feature a distinct set of dance song genres, predominately dal'ouna, 'ataba, and rozana, which vary widely according to local practice. A muṭrib (vocalist) typically begins a dance session with the nonmetered and improvisational vocal genre of mawwal that is crucial for establishing the affective register of the session and engendering ṭarab (literally, 'ecstasy').[6] Instrumentalists traditionally circumambulate the line of dabke dancers, playing the timbrally significant shawm (double-reed pipes indigenous to the area include mijwiz, shababa, and arghul) or ṭabl, a double-barreled drum. At the head of the line of dabke dancers, which can number in the dozens or hundreds, depending on the size of the gathering, the awwal (first) dancer leads a repeated dance sequence with the accompaniment of a supportive dancer, or tānni (second), at their left side. Due to sustained traditions of patriarchy and gendered privilege, the most conspicuous performers at these events are generally men, although women often join the dabke line and may perform more dexterous performances in less public arenas. The gendered dynamics of participation have arguably shifted toward less participation among women due to the reemergence of Islamic groups in the 2000s.[7] Generally, the extent to which dabke as a social dance genre has changed is the subject of much public debate that dwells in the intersection of modernity, morality, and aesthetics.

Dabke is considered khafīf (light) in contrast to the more serious art music practices of the Near East, namely ṭarab and muwashshaḥ (a poetic-musical tradition of Aleppo). The latter are considered the province of sammī'a, or listeners whose carefully honed skills position them as arbiters of musical taste among the cultural elite (Shannon 2006). Many working-class people, however, eschew such distinctions as overrated and prefer not to listen to the "sad" or "heavy" music of ṭarab in everyday life. They prefer the sha'bi (literally, 'of the people') sounds of dabke, a form of street culture that resonates with the popular and populist aesthetics of the Syrian working class.[8] As sound media, dabke can be heard in the transit zones of urban Syria—bus stations, taxi cabs, and radio airwaves.[9] In the 2000s, dabke became heavily commodified as mass media through the proliferation of satellite music videos in this genre. Consumption practices shifted toward pop dabke singers such as Ali El Dik, Wafik Habib, and Fares Karam, who dominated satellite media. These modes of consumption affected local performance economies as those who relied on area gigs for a livelihood were displaced by an uptick in playback media rather than live performance. At the same time, dabke became part of an economy of "distraction" (Benjamin 1969) in the increasingly neoliberal Syria of the late 2000s, a form of audiovisual media that occupied consumer spaces such as chic cafés and shopping malls in urban centers.[10]

The Syrian conflict has deeply affected dabke traditions, as it has every other facet of everyday life. At nonviolent public protests in 2011 and 2012, antiregime activists embraced dabke as a performance technique that embodied collective resistance against the state. Although dabke is affectively associated with weddings, antiregime protestors danced dabke at funerals of the martyred in acts that subverted the affective economy. On social media, dabke proliferated as a protest genre that mobilized the revolutionary public sphere. Singers, dancers, and percussionists became popular heroes. Users mashed up popular dabke tracks with streaming images of the conflict to express their political sentiments, both proregime and antiregime. Others posted documentary clips of dabke at protests, such as those held at Syrian embassies in Europe and Canada, to voice their support from beyond Syria. As with most domains of public life in wartime Syria, dabke was appropriated by all sides of the conflict in ways that attest to its fluidity as a symbol of collective identity.

What has declined in the wake of societal rupture and the weakening of kinship structures is the live performance event at which dabke is practiced. The bleak wartime economy affects disposable income for leisure and entertainment, although levels of nightlife activity vary depending on location, with besieged areas standing in stark contrast to a sustained leisure economy among Damascene elites. There is a decrease in the population of young men due to military conscription, war casualties, and migration, as well as imprisonment and disappearance. This phenomenon not only reduces the number of marriages that occur but has a deep structural impact on kinship relations, which have become significantly fractured. When wedding celebrations do occur in these conditions, the performativity of dabke may suspend social erosion and normalize everyday life, however momentarily.[11] Outside Syria,

*Shayna Silverstein*

dabke remains an indelible part of refugee life for many. Whether and how these performance practices are shifting in relation to migration to Europe, the United States, and Syria's neighboring countries is a topic for future research.

SALṬANA AT THE ḤAFLA

The performance event most associated with dabke is the *ḥafla*—an event of leisure, entertainment, and celebration. Although other kinds of gatherings may occur informally in a private home among friends and family, *ḥaflāt* (plural) typically refer to weddings, shows, and social events hosted at restaurants and nightclubs and in outdoor public spaces. Like other nightlife spaces in which performance is a tactic for "community and agency" (Rivera-Servera 2012: 134), *ḥaflāt* are a key site for the reproduction of social distinctions in Syrian society. Gendered, classed, and racialized distinctions are displayed through the choice of venue, entertainer, and event design, as well as other taste factors such as fashion and bodily comportment. In the mid-2000s, distinctions between Syrian elites and non-elites were differentiated by the display of capital and modes of consumption that shifted due to social market reform.[12] *Ḥaflāt* were integral to cementing those social bonds that ensured access to the privileges and benefits of clientelist networks among the political and business elite.[13] These social events were also vital among non-elites, providing opportunities to celebrate kinship and maintain the cultural capital developed through higher education, vocational training, and other public domains.

*Ḥaflāt* are socially significant as public spaces in which pleasure, desire, and camaraderie engender forms of belonging. People generally mill around; socialize; nibble on appetizers, meats, and sweets; smoke *narghile* (water pipe); drink coffee, beer, or *arak* (anise-flavored liquor); listen to shaʿbi music; and watch or join those on the dance floor. What is remarkable about these events are the ways in which the sound system as a media apparatus affects the sensorium of the ḥafla. Since the transformation of Arab media in the late 1990s, including sound technology, *ḥaflāt* have become increasingly mediatized. At live events, these media-rich, sonically dense spaces are saturated with an acoustic texture of layered, repetitive duple meter rhythms—specifically, the shaʿbi rhythmic patterns of *ayoub*, *nawari*, and *malfouf*. These rhythms are usually produced by a "one-man band" on the keyboard synthesizer, known colloquially as the "org," which also employs presets, samples, and a battery of instrumental timbres (particularly the *mijwiz* and *rebab*). Depending on local

*Disorienting Sounds* [ 247

practices and financial means, other instrumentalists also perform. In particular, ṭabl drummers play a key role through their interactions with participants. Depending on the density of bodies in the performance space, the drummer may roam around and cue participants, specifically the lead dancer, with rhythmic flourishes. In addition, the ṭabl drum infuses the acoustic texture with low-frequency vibrations that impact bodies in the space.

Recent work in sound studies on the impact of haptic and tactile senses on the sonic experience help to explicate the sensibilities of this particular sonic environment. "Haptic" refers to how the vibrational force of sound "impacts the body's flesh, skin, and bones" (Garcia 2015: 64) through the material effects of sonic "transduction," in which sound as a form of energy moves between or across media (see Helmreich 2015). As Luis-Manuel Garcia (2015: 61–63) argues in the context of electronic dance music events, the body's haptic senses are engaged through volume, bass frequencies, and the preponderance of percussive sounds. The impact of sonic texture "across the human body's sensory modes" is such that beats "do not only play an associative or representational role in relation to touch but are impactive and tactile in themselves." Recalling Ahmed's work on spatiality and perception, I might extend this approach to vibrational forces to suggest that those who regularly attend dance music events with booming sound systems tend to become habituated to the ways in which beats shape their movements.

What does this sensory and perceptual approach mean for audile techniques at Syrian ḥaflāt? In contradistinction to the listening practices of ṭarab, which privilege attunement to compositional devices and performance practices as a means of attaining salṭana, sound technology practices at Syrian ḥaflāt make rapt attention to sonic stimuli rather infeasible for nonmusicians. In chatting with other guests and watching the dancers, people opt out of close listening, or focused attention on the aesthetic structures of sound media. This is due in part to transduction techniques at ḥaflāt, which tend to distort as much as they amplify. Events are all evening long: sound systems start at 9 PM and end long past midnight. Duration, loudness, and distortion condition how individuals manage their attention to sound media throughout the event, as loud, distorted, and persistent sounds are emplaced *on* bodies. Rather than a focused attunement to sound aesthetics, audile techniques at ḥaflāt are distracted and inattentive[14] — except when dancing. In light of these audile techniques, it is, I hope, clear that the production of taste and aesthetic pleasures at ḥaflāt is not exclusively oriented to musical performance (although it is certainly dependent on such). Rather, pleasure

*Shayna Silverstein*

is performatively constituted through embodied practices that mediate and are mediated by the multisensorial environment of the ḥafla.

Consider, for instance, the experience of flow during dabke performance. A *dabbik* is a dabke practitioner who displays proficient skill and conveys a feelingful experience on the dance floor; the lead positions in the dabke line are typically occupied by *dabbikūn* (plural). Among dabbikūn, being able to convey a deep understanding of rhythmic aesthetics through movement techniques is critical to the tastemaking and selfhood constituted through dabke performance. Salṭana, in this performance context, registers as rhythmically attuned flow. Salṭana is not only an embodied sensibility, it is also performatively enacted through spoken discourse by which commentators assess if and how a given dabke participant embodies a sense of rhythm. When dabbikūn are "on the beat," they perform ʿāl mazbouṭ, an adverbial phrase that figuratively translates as "with precision" or "in an exact manner." Dabbikūn are praised with phrases such as, "He is precise in dabke"; likewise, modest practitioners may negate or defer their expertise with the phrase, "I don't know [how to do it] exactly." The experience of rhythmically precise flow by a dabke participant, particularly the awwal, also has performative effects on other event participants as they watch, comment, and enjoy these aesthetic pleasures.[15]

To animate the performance dynamic of ʿāl mazbouṭ, I now turn to an description of an improvised session between two practitioners in which salṭana was powerfully experienced by myself and others. This ethnographic encounter stages interactions between a ṭabl drummer and awwal dancer at a ḥaflā that I attended in Lattakia, a city on the northern coast of Syria.[16] Here I describe how salṭana was coconstituted by the dabke dancer and ṭabl drummer as they warmed up the event for the main dance set:

> The awwal dancer raised his arms high, holding a msbaḥ (string of rosary beads) in his right hand and clasping with his left the hand of the second dancer (tānni) in the dabke line. They both shook their shoulders back and forth (hiz). The ṭabl player accelerated the pulse of his drumming as he moved forward to co-perform with the awwal. The muṭrib noticed their interaction and called out to those distracted by table conversation to turn their attention to the dance floor.

> The ṭabl player intensified his playing then eased up, using repeated dum strokes for phrasing. He subdivided strokes by flicking his mallet between low-pitch dum and high-pitch tak strokes. These techniques intensified the rhythmic drive, increasing momentum and propulsion. In response, the

*awwal lowered into a deep knee bend and sustained this low position for*
*an extended moment. He raised his arms high into the air and twisted his*
*knees, still bent, to the right and the left. He moved his right arm in and out,*
*as if waving a sword in slow motion. Then, in a moment of synergetic flow*
*between performers, he looked up at the ṭabl player, who looked right back*
*at him.*

*The org player played a melodic modulation (in the mijwiz preset) that*
*signaled cadential motion toward the end of the mawwal and the beginning*
*of the main dance set. The ṭabl drummer began to anticipate the downbeat*
*cue that he would give to articulate the transition. The awwal dancer antici-*
*pated this key cue with a shallow knee-bend, but maintained his posture as*
*he waited for the next dum stroke. When it arrived, he lowered his knees to*
*the ground and extended his arms higher, his body suspended in anticipa-*
*tion of the resolution of this section.*

*The ṭabl player looked to the other musicians for a cue that would tell*
*him when to transition to the main dance rhythm with his downbeat. In*
*response, the awwal dancer rose from his knee bends and lifted his right heel*
*on the offbeat. He repeated the same motion with his other leg in the second*
*measure. He then anticipated a step exchange from left to right, but, he was*
*early. His eyebrows raised, he acknowledged the misstep by looking directly*
*at the ṭabl player, who returned the gaze, as if they were reassuring each*
*other of recovery and reattunement.*

This moment of disorientation suggests how the aesthetics of rhythmic
precision are actualized in performance. The risk and pleasure of improvised
performance is that one or both might be "off"—by missing a cue or, as in
this case, moving too early. When it did happen, the missed cue arguably did
not affect the experience of salṭana between the drummer and the awwal
dancer. Rather, it heightened their awareness of each other and intensified
their connection (*mittasil*) because it reinforced the emergent conditions of
the performance process. As an ethnographer, I view this missed moment as
a disjuncture in the pleasurable experience of ʿāl mazbouṭ, one that speaks
to the role of multisensory coordination in actualizing salṭana and mittasil
between performers.

In the remainder of this chapter, I describe my first experiences participating in the dabke line at a ḥafla and learning the mechanics of dabke practice. I
problematize how salṭana emerges in particular somatic habitus by describing
how I felt "out of sync" when dancing dabke, that is, not quite able to orient my

*Shayna Silverstein*

body within the overlapping and staggered pulse groupings of the rhythmic and footwork patterns that characterize dabke. This, I argue, is linked to my dance training, the somatic habitus that I bring with me to the field, which shapes the "doing with" of my ethnographic encounters (Madison 2012). Through my own sensory disorientation, I aim to reveal how our perceptual apparatuses tend to direct our bodies in habitual and acquired ways—in ways that demand interrogating the limits of the ethnographer's body in the field.

## DISORIENTATION

At the same wedding ḥafla in Lattakia, I was encouraged to join the dabke line by my host, the brother of the bride. He likely felt that participation would cement my enjoyment of the evening. Our other friends, both sammī'a, declined to dabke with us, giving the excuse that they did not enjoy dancing (a false binary between intellectualism and embodied pleasure often invoked by classically oriented musicians). On my own, I broke into the middle of the dancers, conscientiously choosing to join two women rather than make physical contact with an unfamiliar man. I clasped hands with them, strangers in a room of more than a hundred wedding guests, and began crossing my left foot over my right foot in sync with the *ayoub* and *leff* rhythms of the popular hit "Andak Bahriya" by Wafik Habib.

The iconic gesture of dabke is the foot stomp, for which it is named. It most commonly occurs on the downbeat of the third and final measure of the dabke cycle (a pulse grouping of six), although this position varies widely among the hundred-plus styles of dabke practiced throughout the region. The foot stomp may be performed as a highly exaggerated and intense stomp or a mild-mannered placement of the ball of the foot on the ground, depending on local practice. In this iteration of *dabke al-hourani*, one of the most widely practiced styles across Syria, dancers chose the milder variation. To figure out the beginning of the dabke cycle, I wanted to identify the count on which the foot stomp occurred.

This proved to be a challenge. Although I had not realized it until I joined the dancers, the pulse groupings of the dance and music sequences are not in phase with each other. Dancers tend to dabke in cycles of six pulses, whereas the metrical and melodic phrasing among musicians consists of cycles of four or eight pulses (depending on the *bayt*, or vocal phrasing). This three-to-two tension is somewhat squared up by how the crossover step syncs with the duple meter groove—the left foot falls on the downbeat and the right foot follows on the offbeat. But the foot stomp (along with other idiomatic techniques that vary stylistically) repeats once in each six pulse grouping. This means that while the dance and the music each maintain the same tempo, they have different pulse groupings. The overlapping and staggered dynamics of these pulse groupings challenged my ability to identify the beginning of the pulse grouping in the dance cycle.

To figure out the sequence, I looked at other dancers' feet to watch for the foot stomp. My predisposition toward visual cues rather than aural or proprioceptive perceptions was not incidental. I studied ballet and performed in paraprofessional ballet productions in my teens. The dance ethnographer Deidre Sklar (2000: 72) captures my experience when she writes, "In the world of studios and rehearsal halls we learn to translate visual and verbal information into movement sensation. A teacher demonstrates a position: we see it and 'try it on' in kinesthetic imagination." My ballet classes had engrained this particular flow of visual, verbal, and kinesthetic techniques in me from a young age. In the studio I learned to quickly regurgitate the patterns, sequences, and choreographies that were visually and verbally cued to us as young dancers. As Pierre Bourdieu (1977) has observed, embodied practices mediate and reproduce individual's positions with a given social space, or habitus. Social classes reproduce their benefits and privileges from generation to generation through the cultivation of bodily disposition in everyday actions that, as Bourdieu has argued, mask the very socialities that they sustain. After years spent memorizing ballet sequences transmitted through my teachers' visual and verbal cues, I rehearsed these techniques when looking around for visual cues on the dance floor of the ḥafla. In this moment of encounter between different movement worlds, my entrained disposition as a ballet dancer shaped how I perceived the gestures and movements that constituted the socio-kinetic space of this wedding. In particular, I privileged the visual mode of perception, understanding it as a self-evident tactic for disciplining my body in a dance world.

However, as I learned from subsequent conversations about dabke technique and disposition, skilled dabke dancers do not depend exclusively on visuality, or even audibility, for orientation. Although casual participants may occasionally look around, dabbikūn depend on one another in a specific way, akin to "feeling oneself to be in the other's body moving" (Sklar 2000; cf. Albright 2013: 11). Sklar terms this intersubjectivity "empathic kinesthetic perception," which she describes as a combination of mimesis and empathy that "implies a bridging between subjectivities, a 'connected knowing' that produces a very intimate kind of knowledge . . . temporary joining in empathy produces not a blurry merger but an articulated perception of differ-

*Shayna Silverstein*

ence" (Sklar 2001b; cf. Albright 2013: 11). This connectedness is, I suggest, central to the experience of salṭana among dabke practitioners.

As I demonstrate in the next encounter, this connectedness also challenged another of my predisposed notions of selfhood cultivated by my ballet training: that of the autonomous body on the dance floor. Here I discuss how learning to anchor the awwal dancer helped me to better understand the kinesthetic self as "an interdependent part that flows through and with the world" (Albright 2013: 293). Privileging the historically contingent body as a site for the contestation of socialities, the next section demonstrates how fieldwork emerges through corporeally, sensorially, and relationally invested interactions with interlocutors that challenge me as an ethnographer to "re-create" myself on the dance floor.

### FINDING BALANCE

In Syrian dabke, Sklar's concept of "empathic kinesthetic perception" between performers best translates into the experience of mizān (balance). This term conveys the balance of motion between performers on which each depends to be ʿāl mazbouṭ. Performers insist on feeling the weight of the other to better perceive and situate their actions in the course of performance. These dynamics are vital for the performance relationship between the awwal and tānni lead dancers, which engenders salṭana. Or, as my teachers insisted, they would be better able to perform variations and improvise more skillfully if I maintained the fixed-step sequence with certainty and directionality.

As I mentioned in the opening of this chapter, one learning session in central Damascus fell apart when the young dabbik became disoriented by my inability to manage the role of tānni. In a session with another dabbik in Aleppo, Ahmad also tried to lead while I practiced dabke alongside him. Yet we both became disoriented when we could not figure out how to sync our dance movements. Ahmad was an Arabic teacher for foreign students, and he drew on these teaching skills to overcome this challenge. He asked me to "anchor" him by reinforcing the duple rhythm with crossover steps. We resumed our dance session. I thought I was anchoring well, but Ahmad said that he needed more from me to feel mizān. I responded with enhanced proprioception—my handgrip, length of my arm, shoulder attachment—on my right side, which connected me with Ahmad. Although I drew on my dance experience to engage proprioception, this was a meaningful shift from the solo, autonomous, and individuated movements in dance studios to which I was accustomed. Learning how to "anchor the awwal," as Ahmad suggested, enhanced my awareness of kinesthetic and proprioceptive perception rather than visual or auditory perception. The performative effects of this were not only that my body learned how to be a responsive and supportive tānni such that the lead dancer could embellish, individuate, stylize, and improvise, but that I finally experienced salṭana in a dabke session.

Becoming a dabbik depends on a balance constituted through performance with others. The vitality of this performance as ethnographic encounter is in how it shaped our relationship to each other as teacher and student, cultural insider and outsider, male and female, experienced dabbik and trained ballet dancer learning a new movement tradition. Being disoriented was generative in the sense that it destabilized the performative effect of bodily acts that become naturalized through their reiteration over time, or what Ahmed pinpoints as habitually directed movements through which bodies acquire their shape. In this coperformative moment, Ahmad and I "bridged" these subjectivities through acts of "empathic kinesthetic perception." Through imitation and empathy we were able to work through sensory disorientation and achieve "connected knowing"—that is, we transmitted embodied knowledge through a shared sense of mittasil that emerged through the pedagogical process. It was not only the coevality of a shared practice-based experience that generated salṭana in this ethnographic encounter but also the process of transmitting embodied knowledge and experiencing mizān.

It is difficult to underestimate the value of becoming a dabbik for young Syrian men. The awwal is a critical role on the dance floor as this person intensifies and heightens performance at a ḥafla. In contemporary Syrian society, the awwal is almost always male. He may be a person of distinction at the ḥafla, such as the father of the bride, the eldest male relative, or a young cousin or neighbor who is particularly gifted at dabke and can maintain social prestige through performance. Women may also perform the role of awwal, but this tends to occur more often at all-female gatherings and socially progressive venues. In our dabke session, Ahmad shared that he was known among his peers as a dabbik "who always takes the lead" at gatherings. Which steps and embellishments he performed at a given event, even the extent to which he was rhythmically precise as awwal, were often incidental to his masculinized acts of assertion among his peers. Like other domains in which Arab masculinity is engendered through relationships with others, particularly female kin, the performance of mizān conveys how the emergence of (mostly) masculine selfhood through dabke is constituted through the empathic kinesthetic perception of others. These performance

*Shayna Silverstein*

practices are vital for the production of masculinity among young men who are challenged by the myriad socioeconomic issues Syrians faced in the 2000s and continue to face today.[17]

Throughout this essay, I have deployed disorientation as an ethnographic strategy that dehabituates modes of sensory perception and opens up new possibilities for understanding how bodily techniques shape and are shaped by social relations in field sites. I have described how particular moments of disorientation illuminated my sensory disposition, such as a proclivity for visual orientation while dancing and the autonomy of the individual body on the dance floor. I was also disoriented by the rhythmic dynamics of dabke, in which music and dance cycles tend to be out of phase with each other. These experiences of rhythmic, kinesthetic, and proprioceptive disorientation became welcome opportunities to better understand performance dynamics in Syrian popular culture and how salṭana is achieved through the particular ways in which people dabke together to shaʿbi dance music. Furthermore, the sociocultural intimacy of ḥaflāt engendered through the embodied dynamics of dabke practice specifically occurs through a sense of balance between dabke practitioners and the display of rhythmic precision among dancers who play with and against the pulse groupings of music and dance cycles. What sensory disorientation suggests is that participation at ḥaflāt is about much more than listening. Although sound systems dominate the aural environment in ways that necessitate techniques of distracted and inattentive listening, when event participants do engage attentively with the sound system, it is through kinesthetics and collective movement.

Another way to approach these dynamics is to suggest that listening is a process that occurs with the whole body, rather than solely with the ear. Charles Hirschkind (2006) suggests as much in his ethnography of contemporary cassette culture among Islamic followers in Cairo. Among cab drivers and others negotiating the urban environment of Cairo, the performative act of listening to sermons, recorded and live, embodies piety. More than an aural engagement with sound media, listeners are those who "open" their "heart" and enact a repertoire of whole-body gestures in response to religious prayers and speech idioms. Listening attentively to sermons on cassette tapes cultivates an ethical space that is formed through exclusion of and disengagement from the urban environment—in other words, listening is a mode of attention and pious intention in a political economy of distraction. This engagement

*Disorienting Sounds*                                        [ 255

with listening as an embodied experience is strikingly different from how listening is generally approached by the field of sound studies, which tends to posit the ear "as a synecdoche for the whole of the listening body" (Kheshti 2015: 56). For example, in his cultural history of sound media in America, Sterne (2003) details how scientific curiosity about deafness fostered the development of the first sound recording technology, an instrument that reproduced the vibrational effects of soundwaves on the ear. These inventions, and their inscribed role in modern recording history, have contributed to the reification of the ear among culture industries and academe. As Roshanak Kheshti (2015: 55) writes, the ear has become a "body part that is the structuring logic for how aural fantasies function." The challenge for sound studies is how to account for the reification of the ear, and attendant processes of hearing, listening, and audibility, while disabling the production of aural fantasies that arguably motivates much of the field.

This essay seeks to avoid this direction by detailing how, to echo again Ahmed's line of critique, bodies are "directed" in sonically dense environments. Movement is crucial to the lived experience of the ḥafla. The assemblage of dozens of dancers repeating the same patterns in coordinated movement demonstrates how bodies acquire directionality and shape through cross-modal perceptions and interactions with sounds, bodies, and objects. Consideration of somatic, spatial, and proprioceptive dynamics, as well as material tactility, expands our sensory approaches to sound beyond that of aurality. Moreover, by remapping or redistributing sound studies throughout the body, we might think of bodies as "zones of varying intensity that respond differently to stimuli, such as proximity, heat, light, and movement."[18] Ḥaflāt in Syria are exemplary spaces for tackling this line of inquiry insofar as the wide variety of activities taken up by event participants at any given moment during the evening-long sensorium suggest how different zones of the body are stimulated and actively engaged. Finally, what this suggests is that how people engage with sound—whether aurality, audibility, or attentive listening—is diffracted by what other activities are also at play. It is not that listening is singular, but whether and how listening practices produce cultural meaning in particular social settings.

This essay has focused on disorientation produced through ethnographic encounter to bring attention to the social relations that are constituted through fieldwork and the implications of these encounters for the field of sound studies. As an ethnographic strategy, sensory disorientation not only helps to shed light on the mechanics, techniques, and lived experience of cultural practices such as dabke; it also recognizes the role of embodied practices in "constitut-

*Shayna Silverstein*

ing knowledge, emotion, and creation" (Madison 2011: 185). What embodied ethnographic encounters offer is that sensory and sound studies are not only about our individual bodies but about the intersubjectivity of bodies interacting with one another. Rather than imposing the question of the senses on Others in ways that make them carry the burden of nonvisuality, I suggest that we reflect back on ourselves to better understand the effects that are produced by the dominance of visuality and audibility, especially in our encounters in the field. How we listen with others listening, and how we dance with others dancing, begins with disorienting audibility as the primary perceptual mode for sound studies and striving for somatic modes of attention that articulate our mutually constituted somatic-sonic imaginaries of the field.

## Notes

1. This name and all other names are pseudonyms.

2. For ethnographic studies of ṭarab, see Bordelon 2011; El Zein 2016; Racy 2003; Shannon 2006.

3. This is a reference to Kheshti's (2015) critique of listening in which she argues for the ear not as a reified organ for hearing but as a signifier of how modern listeners engender aural others.

4. Tim Ingold (2000) argues that the dialectic of sight and sound dominates the disciplinary agenda of sound studies. He likens the totality of the senses to a "whole cake" rather than dividing the senses into "slices of cake." For instance, blind or deaf people are not missing a slice of the sensory cake; rather, for them the cake is smaller. Ingold also encourages relationality among the senses—for instance, how the absence of a sense, such as vision, reveals features and properties of aural perception, among both blind and sighted people, that otherwise would be taken for granted.

5. For a historical ethnography of Syrian dabke, see Silverstein 2012.

6. In colloquial and academic practice, the term 'ṭarab' is used interchangeably to refer to either an affective state or a historically specific genre of art music associated with modern Arab culture.

7. Ahmed Sadiddin, personal communication with author, August 20, 2016. See also Khatib (2011) for her study of Islamicization and piety practices in twenty-first century Syria.

8. For a discussion of class dynamics and popular music in Syria, see Silverstein 2012.

9. For an ethnography of Syrian radio, see Bothwell 2013.

10. For essays on Syrian politics, society, and culture during Assad's first decade of rule, see Salamandra and Stenberg 2015.

11. See Declan Walsh, "On the Ground in Aleppo: Bloodshed, Misery and Hope," *New York Times*, April 30, 2016.

12. For an ethnography of consumption and class in Damascus, see Salamandra 2004.

13. For relations between the state and business elite (1970–2005), see Haddad 2011.

14. Larkin (2014) conceptualizes "techniques of inattention" as practices through which individuals negotiate the impact of the media object (such as a loudspeaker) on the human body, particularly in relation to urban sounds in public spaces.

15. This is similar to how audiences of ṭarab music feel "ecstasy" or "enchantment" when listening to a musician who achieves salṭana.

16. At this time, Lattakia flourished through regional trade and commerce with Lebanon and Turkey, as well as political and economic ties to Damascus through Alawi networks. Though considered the Alawi heartland, the area has diverse constituencies, including Sunni and Christian elite.

17. The "crisis of masculinity" faced by young Syrian men in the 2000s has generally been attributed to lack of social mobility and a highly educated labor force facing high rates of unemployment, among other factors.

18. I thank Gavin Steingo for this point.

## References

Ahmed, Sara. 2006. *Queer Phenomenology: Orientations, Objects, Others*. Durham, NC: Duke University Press.

Albright, Ann Cooper. 2013. *Engaging Bodies: The Politics and Poetics of Corporeality*. Middletown, CT: Wesleyan University Press.

Benjamin, Walter. 1969. "The Work of Art in an Age of Mechanical Reproduction." In *Illuminations*, ed. and introd. Hannah Arendt, trans. Harry Zohn, 217–52. New York: Schocken Books.

Bordelon, Candace Ann. 2011. "Finding the 'Feeling' through Movement and Music: An Exploration of *Ṭarab* in Oriental Dance." PhD diss., Texas Women's University, Denton.

Bothwell, Beau. 2013. "Song, State, Sawa: Music and Political Radio between the US and Syria." PhD diss., Columbia University, New York.

Bourdieu, Pierre. 1977. *Outline of a Theory of Practice*. Cambridge: Cambridge University Press.

Csikszentmihalyi, Mihaly. 2008. *Flow: The Psychology of Optimal Experience*. New York: Harper Perennial.

Csordas, Thomas J. 1993. "Somatic Modes of Attention." *Cultural Anthropology* 8, no. 2: 135–56.

El Zein, Rayya S. 2016. "Performing el Rap el 'Arabi 2005–2015: Feeling Politics amid Neoliberal Incursions in Ramallah, Amman, and Beirut." PhD diss., City University of New York, New York.

Garcia, Luis-Manuel. 2015. "Beats, Flesh, and Grain: Sonic Tactility and Affect in Electronic Dance Music." *Sound Studies* 1, no. 1: 59–76.

Haddad, Bassam. 2011. *Business Networks in Syria: The Political Economy of Authoritarian Resilience*. Palo Alto, CA: Stanford University Press.

*Shayna Silverstein*

Helmreich, Stefan. 2015. "Transduction." In *Keywords in Sound*, ed. David Novak and Matt Sakakeeny, 222–31. Durham, NC: Duke University Press.

Hirschkind, Charles. 2006. *The Ethical Soundscape: Cassette Sermons and Islamic Counterpublics*. New York: Columbia University Press.

Ingold, Tim. 2000. "Stop, Look and Listen! Vision, Hearing and Human Movement." In *The Perception of the Environment: Essays on Livelihood, Dwelling, and Skill*, 243–87. London: Routledge.

Kapchan, Deborah. 2017. *Theorizing Sound Writing*. Middletown, CT: Wesleyan University Press.

Khatib, Line. 2011. *Islamic Revivalism in Syria*. Abingdon, UK: Routledge.

Kheshti, Roshanak. 2015. *Modernity's Ear: Listening to Race and Gender in World Music*. New York: New York University Press.

Larkin, Brian. 2014. "Techniques of Inattention: The Mediality of Loudspeakers in Nigeria." *Anthropological Quarterly* 87, no. 4: 989–1016.

Madison, D. Soyini. 2012. *Critical Ethnography: Method, Ethics, and Performance*, 2nd ed. Thousand Oaks, CA: Sage.

Peake, Bryce. 2016. "Noise: A Historical Ethnography of Listening, Masculinity, and Media Technology in British Gibraltar, 1940–2013." *Cultural Studies* 30, no. 1: 78–105.

Racy, Ali Jihad. 2003. *Making Music in the Arab World: The Culture and Artistry of Ṭarab*. Cambridge: Cambridge University Press.

Rivera-Servera, Ramón H. 2012. *Performing Queer Latinidad: Dance, Sexuality, Politics*. Ann Arbor: University of Michigan Press.

Salamandra, Christa. 2004. *A New Old Damascus: Authenticity and Distinction in Urban Syria*. Bloomington: Indiana University Press.

Salamandra, Christa, and Leif Stenberg, eds. 2015. *Syria from Reform to Revolt: Culture, Society, and Religion*. Syracuse, NY: Syracuse University Press.

Shannon, Jonathan. 2006. *Under the Jasmine Trees: Music and Modernity in Syria*. Middletown, CT: Wesleyan University Press.

Silverstein, Shayna. 2012. "Mobilizing Bodies in Syria: *Dabke*, Popular Culture, and the Politics of Belonging." PhD diss., University of Chicago.

Sklar, Deidre. 2000. "Reprise: On Dance Ethnography." *Dance Research Journal* 32, no. 1: 70–77.

Sklar, Deidre. 2001a. *Dancing Like a Virgin: Body and Faith in the Fiesta of Tortugas, New Mexico*. Berkeley: University of California Press.

Sklar, Deidre. 2001b. "Five Premises for a Culturally Sensitive Approach to Dance." In *Moving History/Dancing Cultures*, ed. Ann Cooper Albright and Ann Dils, 30–32. Middletown, CT: Wesleyan University Press.

Sterne, Jonathan. 2003. *The Auditory Past: Cultural Origins of Sound Reproduction*. Durham, NC: Duke University Press.

Sterne, Jonathan. 2012. *The Sound Studies Reader*. New York: Routledge.

# Afterword

## SONIC CARTOGRAPHIES

*Ana María Ochoa Gautier*

.....

Damascus, Buenaventura, Bangalore, Bangkok, the Marshall Islands, KwaZulu-Natal, the Colombian AfroPacific, the French *banlieus*, Soweto, Marshallese communities in Arkansas, Sri Lanka, Brooklyn. . . . The theoretical importance of the cartographic intention at the heart of remapping sound studies emerges clearly through a simple enumeration of the places in and from which the sounds in this volume are thought, heard, perceived, and felt. This apparently simple gesture of radically altering the places from which a corpus of works is assembled enacts a profound theoretical and acoustic displacement because it makes us realize how sonically marginal these places are to sound studies. Herein lies the reason behind this volume: it problematizes the politics of sound studies by calling attention to the logic that simultaneously links sound and South as inextricable to the history of the colonial modern, as well as to a contemporary elision of sound practices and sound theories from the South brought about by the normativization of the field provoked by its disciplinization. The paradox is that sound studies has enacted these epistemic relations while pretending not to do so.

In this volume, reference is frequently made in different chapters to key works in sound studies that consciously articulate sound's problematic relation to the modern. In that sense, the chapters' authors signal how sound studies tends to reproduce a classical relation between knowledge and modernity that has been denounced by postcolonial and decolonial discourses for decades. This relation structures an epistemic tendency to universalize a particular experience as speaking for all, while ascribing particular modes of knowledge- and sense-making in sound to the South. While enacting this critique, the authors in the volume simultaneously recognize key contributions to thinking sound and South. So the paradox that this book makes evident is that sound studies reproduces a modernist-colonial slant smack in the middle of times of postcolonial critique and decolonial deconstruction and as a discipline that, in many of its key writings, emerges as critical of modernity.

The volume's point of departure is twofold. First, it makes explicit the recognition that sound studies has become an established, institutionalized discipline. This is a classical colonial-modern critique, and in that sense the volume builds on previous works in sound studies that have articulated this. The editors suggest some of these works in the introduction, and the chapters' authors reference them. Second, the volume's objective is not to simply expand the archive of sound studies comparatively to diversify the North via sound studies in the South. It does not—to draw an analogy—attempt a "world music" survey of sound that would simply keep the status quo of what is meant by sound in sound studies by othering alternative presences that have not had theoretical primacy in the recent disciplinization of the field. The point here is to provoke a conceptual and political unsettling of the field.

Yet the critique of a discipline that consolidated as such in the late twentieth century and early twenty-first century does not happen under the same political or institutional conditions of disciplines that consolidated in the late nineteenth century and that have been the primary object of postcolonial and decolonial critique. Today, the inauguration of sound studies as a recognized field and discipline of study happens through an imperial politics of finance that shapes the very academic institutions that allow the North to name, apparently for all, a phenomenon that arguably is taking place worldwide. Sound studies has consolidated itself as a recognized discipline amid a neoliberalization of the politics of academic production that tie it to an "audit regime" characteristic of recent modes of managerial internationalization of disciplined knowledge worldwide (Strathern 2000). In such regimes, academic production and reproduction are characterized by a "culture of accountability" that intertwines "twinned precepts of economic efficiency and good practice" (Strathern 2000: 2) with a culture of management that deeply neutralizes or antagonizes, "in a language often difficult to recognize," the politics of exclusion that such regimes articulate (Strathern 2000: 2).

Particularly complex is how languages of "moral responsibility" and cultural critique embedded in such an academic structure are neutralized by the managerial and economic regimes of academic efficiency through which

*Ana María Ochoa Gautier*

they are deployed (Richard 2001; Strathern 2000). Sound studies as a named, recognized discipline appears via a publishing industry located in the North and a globally articulated, unequal politics of recognition and citation enacted by the labor politics of the scientific and publishing institutions and structures in which we work and through which we disseminate our knowledge. So sound studies' particular form of disciplinary consolidation articulates the contradictions between the critique of colonial modernity and the deployment of neoliberal efficacy in the way its short history neutralizes the polemical relation between South and sound in its discourse and in the way it privileges a particular epistemic experience of colleagues located in the North as standing for sound studies. For example, even though they have been working in fields akin to sound studies for decades, colleagues working in Latin America and the Caribbean are forced, through the global intellectual politics of citation and recognition, to speak of sound studies as an a priori *named by* and *emerging from* the North (Iazzetta 2015).

Perhaps it is no coincidence, as the editors of the volume point out, that one of the significant elements missing from Jonathan Sterne's (2003) much cited "audiovisual litany" is precisely the relation between South and sound. It simply appears embedded in all of the other items of the litany but—in not being explicitly acknowledged—emerges as the central spectral presence that animates its discourse. This is so precisely because its spectrality is not only discursive and is not merely related to the political oral technology of Christianity (Sterne 2003). It is also articulated through the contemporary imperial, neoliberal, institutional structure of academic production, recognition, and citation that articulates the global governance of the capacity to *name* the emergence of a field. The privileged genealogy of sound studies, to be sure, has been challenged, almost since the very emergence of the field (see for example Campos Fonseca 2014; Erlmann 2004; Estévez Trujillo 2008, Feld 2012b; Ochoa Gautier 2014; Szendy 2015). But this does not alter the fact that sound studies becomes entrapped not only by its own specific exclusions (Steingo 2016), but also by the way discourse is embedded in the global politics of academic citation, labor, and finance. Moreover, despite challenges to its genealogies, modes of thinking about the field that emerged simultaneously and involved challenges to the mode of narrating what counts as a proper sound as articulated by experiences in the South are conspicuously absent from the description of its disciplinary formation.[1]

What this book points out is that, because sound studies has been presented as a highly dispersed field that has emerged slowly over the past century (Hilmes 2005; Sterne 2012), its coalescing into a discipline as a coming together of an erratic multiplicity that is critical of modernity erases the very hierarchies from which the field operates—or, at the very least, it potentiates such erasures. So one of the key problematics of sound studies today is how the contention that it challenges the privilege of music and language as primary and preeminent fields of the sonic (Smith 2004), and the fact that its origins are disciplinarily multiple (Sterne 2012), can now produce or perform generalizations (or origin narratives) that have the potential effect of enacting a depoliticization of the field of sound studies itself, by eliding the very multiplicity it purportedly embraces through a homogenization of its key theoretical terms.

As a response to this situation, this volume seeks to remap sound studies through several strategies. First, it calls attention to how key words in sound studies—"listening," "deafness," "silence," "noise," "voice," "sound technologies," to name just some of the classical terms that are central to the definition of the field—emerge with different genealogies, embodiments, manifestations, and theorizations when thought from different experiences of the world in the South. Second, it highlights how different sounds, forms of listening, and uses of technology that are central to many people in the world (perhaps the majority) remain confined to the parameters of "case studies" by their ethnographic specificity. They are understood as terms meant to display the Other, and, as such, have not been taken into account by the discipline in setting up the very parameters of its theoretical domain. This book responds to this situation through a focus on ethnography rather than historiography, which until now has been more central to the field (Erlmann 2004; Porcello et al. 2010). The problematic of the interpretation of the sound archive and its anthropological and historical renderings become central, then, to the theoretical questions that are articulated here.

EQUIVOCATION AND THE THRESHOLDS OF THE SOUTH

Equivocation, say the editors, runs at the center of this editorial labor. Let us, in closing this volume, reproduce these words from the introduction because of the theoretical centrality in the interpretive apparatus the editors propose: "Remapping sound studies participates in a remapping—and, indeed, a partial decolonization—of thinking and listening. Drawing on Viveiros de Castro's (2004) notion of conceptual 'equivocation,' we advocate a 'transformation or even disfiguration' (Holbraad et al. 2013) of thinking about sound, about ways of hearing, and about the constitution of entities that hear via Southern perspectives. To 'remap' sound studies, then, means engaging po-

*Ana María Ochoa Gautier*

tential equivocations head-on—listening across time and place in a manner that lives up to the challenges of twenty-first-century geopolitics."

Indeed, conceptual equivocation has already contributed key critical labor to, and is an important theoretical trope in, key works in sound studies that question the relation between the anthropological, the colonial, and sound (Helmreich 2007; Ochoa Gautier 2014). Thus, the authors build on existing theoretical labor. In calling forth "conceptual equivocation," what are the editors of this volume advocating? What do they mean by inviting readers to enact a transformation, or even a disfiguration, of thinking about sound, about ways of hearing, and about the constitution of entities that hear via Southern perspectives?

In the central role it gives ethnography in articulating sound and the South, this book raises the question of the status of what historically has been considered the anthropological archive and, more specifically, of the anthropology of sound to sound studies. One could say that ethnography has long ceased to be a method used only by anthropology. But a method is not a discipline. The confusion of anthropology with the method of ethnography may be one of the reasons that sound studies has not fully embraced its proximity to "doing anthropology in sound" (Feld and Brenneis 2004). Thus, in this volume we find an unresolved threshold between anthropology in sound and sound studies that is made evident by the diversity of anthropological approaches to sound ethnography found in this volume. Indeed, the proposal of conceptual equivocation is one dimension in which such a relation has begun to be done. But do all of the chapters engage the proposal of equivocation suggested in the introduction? They do not. One of the characteristics of this book is the diversity of anthropological approaches to the sound objects themselves. I point this out simply to distinguish between the desired direction of the editors, expressed also by some of the authors in some of the chapters, and the diversity of anthropological theories in sound that appear here. This disjuncture names precisely the threshold this book opens up: a feedback between the notion of anthropology and the anthropology of sound as mutually constituted spheres that deserves a further theoretical elaboration of its multiple dimensions.

By "threshold" I do not mean a boundary with another field but, rather, the alteration of substance implied by acknowledging difference and the work of comparison and transformation it generates not by generating a new margin, but as central to thought. With regard to this, I raise three interrelated issues. The first that emerges as one reads the essays in the volume is the question of the relation between anthropology of sound and its closely related anthropological fields (e.g., ethnomusicology, sociolinguistic anthropology, sensory anthropology, media anthropology, acoustemology, and so on) in framing sound studies. The second has to do with the relation between equivocation and "remapping" the field. The third, derived from the previous two, has to do with the specific politics of the decolonial that the volume raises, since this is a book not about sound and coloniality but about *remapping* sound studies. Thus, the politicization of cartography through sound implies the centrality of the decolonial as a politics but not the subsuming of all of the political implications of the verb "remapping" under the figure of the colonial. Among these three issues, a central problematic emerges about the politics of displacement of sound studies and about the thresholds and routes it opens up. In what follows, I do not take up these three issues separately but address them interrelatedly.

By highlighting equivocation, the editors draw attention to an anthropological debate that recasts the centrality of the conceptual in the work of the decolonial (Viveiros de Castro 2010). As stated earlier, the problem of calling attention to the South in sound studies is that it unsettles the very way we think the key terms of the sonorous via different experiences of the South.

To think via Southern perspectives in order to live up to the geopolitical challenges of the twenty-first century implies giving political importance to thinking itself—not in the solipsism of academic self-immersion, but in the thresholds of a world that cannot give itself the luxury of denying that we are living at the crossroads of the South. This implies, for example, attending to forms of "crisscrossing" as a way to listen (chapter 2) and to "sounding out" as a means of queering conventions that sequester us in the sonic prevalence of sounds considered virtuosic (chapter 7). It also implies the theoretical elaboration of ideas about movement and sound that are proposed here as concepts of "navigation" between sites of voicing (chapter 3), of "turning back" interpretations between experiences of sound and using such turns as theoretical frames (chapter 9), of "disorientation" as a site of conceptual formation (chapter 11). We also find crucial the aural relation between the human and nonhuman (chapter 4), the polity and politics of loudness (chapter 5), forms and uses of the technological that alter its very conceptual ground (chapter 1), and the challenges of thinking through the comparative politics of religion (chapter 9). These different dimensions emerge in different chapters of this book.

Based on this, one of the key contributions of this volume is the theoretical potential that the concepts that emerge throughout the chapters hold for unsettling the premises of the sonorous in sound studies. If billions of

*Ana María Ochoa Gautier*

women ululate, as Louise Meintjes says, then how would taking this as a conceptual problem, and not as just one more case study from the South, unsettle the framework for thinking the grammatology of the voice? If the value of hearing and of the aural is modulated by different people according to specific circumstances, as Jairo Moreno, Michele Friedner, and Benjamin Tausig propose, then how does such a contention imply rethinking the histories of deafness, hearing, and technology that have been described in much of the mainstrem literature of sound studies as "globally" significant? Many more questions can thus be articulated when one reads this book. The chapters in the volume chart routes for how that can take place.

In this sense, the volume invokes other texts that advocate thinking at the junctures of anthropology and sound. The parallelism between the quote on equivocation from the editors of this volume that opens this section and one in which Stefan Helmreich calls for "transductive thinking" that, via sounded anthropologies, critiques the anthropological field itself is telling and important. He writes, "Rather than thinking immersively or reflexively, then, what about thinking transductively?" Extending the chain of those who have thought transduction, Helmreich cites Adrian Mackenzie, who in his turn had built on Simondon's work, and continues: "To think transductively is to mediate between different orders, to place heterogeneous realities in contact, and to become something different" (2002: 18). To think transductively is to attend to the earache, the imbalance (Helmreich 2007: 633). I attend to the parallelism between transformation and disfiguration, in the quote by Steingo and Sykes, and to earache and imbalance, in the one by Helmreich, in order to highlight the chain of thought that links borderline conceptual transformations that are characteristic of the way the South makes us rethink the whole conceptual field. Perhaps one of the reasons the editors chose equivocation instead of transduction is that equivocation, as stated by Steingo, is a form of resistance to the frequent, facile association between transduction and transmission (Steingo 2016) or indeed to highlight the fact that not all equivocation ends in transformation (as is implied by transduction). Keeping the disjuncture between equivocation and transduction alerts us to the implications of the type of comparative thinking advocated by an ethnographic decolonial link between sound and South. This is not just a turn to anthropology. It is a turn to a type of anthropological work that in itself implies a transformation in thinking. What is advocated here is not an "ethnosoundology" that acknowledges that all fields of sonic thought are simultaneously constituted by culture, such that it is important to attend to cultural specificities of different peoples, places, and histories in sound. The conceptual transformation that is invoked here is one that radically shifts the conceptual labor of thinking sound—by questioning the very ground of thinking such key terms as "culture," "nature," "technology," and their relation to key terms in sound studies mentioned above—through thinking the South. The keyword here, then, is "and" (Viveiros de Castro [2003] 2015) and the operation one of "partial connections" (Strathern 2004).

So, does this imply that we are discarding ethnographic specificity in the service of conceptual elaboration, as some have argued in the critique against an anthropology that calls for a recasting of the conceptual ground of the field? Not at all. If anything binds these very different essays, it is the way empirical ethnographic work is the site of generation of the conceptual labor that requires a transformation in thinking about sound through elements that emerge in comparison. These, in turn, unsettle the very premises on which such thinking is based. "South" unsettles sound studies in that it questions its intention as an accumulation of anything and everybody that deals with sound, and reorients it to the question of what is meant by sound as one that always involves a theoretical operation of comparison and transformation—in short, of bricolage in the disjunction between equivocation and transduction. Equivocation always implies more than one. It always happens between the two and its many, even when the two resides in the very concept of the one (Strathern 1988; Stolze Lima 1999). But it also implies that rather than find a general principle to bind the theories from and about the South or a prevalent mode of analysis that supersedes others, the configurations that emerge in rethinking South and sound are always partial connections rather than new paradigms, genealogies, or the search for overarching concepts (Strathern 2004).

How to frame the significance of highlighting the relation between the anthropology of sound and the relation between sound and South? Much work in the anthropology of sound is not solely about responding to ocularcentrism in sound studies. Nor do the works of sound studies and anthropology of sound, by calling attention to the entanglement of colonial power, ritual, sound, and the problems of political theology simply unquestioningly reproduce a critique of political theology leaving untouched the foundational concepts of comparative religion.[2] So what is at stake here? By highlighting equivocation as one of the potential partial connections through which to trace a conceptual route of thinking sound and South, the editors of this volume emphasize that such a route requires patient theoretical and intensely comparative work.

*Ana María Ochoa Gautier*

Thus, naming a lack, as in the lack of South in sound studies, is not about offering a new genealogy or about creating another "auditory (re)turn" that presents as new that which has already "been practiced in theory" (Szendy 2015), even if marginalized by sound studies. Instead, the editors of this volume seek to provoke a change of route by the very act of naming. Let us not forget the paradox of that act of naming: it is simultaneously that which invokes what needs to be named in order to denounce and that which is profoundly modernist in calling attention to its very formation. Perhaps that is why the call to attention in this volume is especially about the very terms of audibility—as is emphasized repeatedly throughout the chapters—and the people that speak through them: listen to us; listen to how we listen. Such a politics of the relation among the named, the audible, and equivocation invites the possibility of seeking the confluence or companionship of routes that also register the political effects of the tension between the need to denounce and the modern-colonial politics that hijacks the language and formats of recognition. Thus, one finds that this volume opens up a need for dialogue with feminist theory (Grosz 1999; Haraway 2016), queer theory (Halberstam 2011), and Black studies (Moten 2003; Weheliye 2014), as well as with other theoretical approaches that question the relation between ontology and epistemology in such a way that naming is not confused with *inaugurating* a topic but, rather, is an event that *enables* a materialization that emerges through transformation and partial connections.

The important claim here is not whether this volume remaps sound studies understood as an accomplished event. It is the cartographic event that names and materializes the threshold that opens up the navigational routes that are possible in the naming of the intellectual labor implied. Thus, the relation between threshold and excess is deeply embedded in the question that the relation between sound and South has always raised, because one of its central political issues is about the relation between the human and the nonhuman. This text does not advocate turning to ontology and thereby depoliticizing the very question about life in the recognition of the singular. It seeks, rather, to overturn the relation among the ontological, the sonic, and the political via the lives and afterlives of the sonic in the South and the modes of that which constitute their acoustic assemblage.

Here the ontological question appears not as a displacement of other problematic sonicities that are key to thinking early twentieth-century relations between ontology and sound (see, e.g., Maniglier 2006). It is not meant either as an elision of the problematics of the political. Rather, it highlights how questions about forms of life and forms of death have always been central to how sound and South are entangled. Thus, it is also important to understand that what is advocated here is not a so-called turn to an emphasis on the ontology of sound studies via the South but, rather, an alteration of the very grounding of the relation between nature and culture, between technique and technology, and between the human and the nonhuman required by the politics of life and death in the South. It is also a call to figure how the parameters by which the call to forms of life and death in sound are themselves attended to by the political conditions in which listening and producing sound always happen.

A Benjaminian question about the *how* of citation is raised in this volume. In calling attention to the South, this volume is not solely about denouncing the seeming cyclical return of the classical relation between metropole and periphery repeatedly rehearsed in the history of colonial critiques of various sorts: while scholars from the South by necessity cite their experience and that of the center in their formations of knowledge, the same necessity is not true of the center in casting its intellectual disposition. Rather, the texts in this volume speak of a global condition of intellectual displacement provoked by attention to the South. It is perhaps not by chance that at least half of the chapters' authors are migrants leading lives between worlds of belonging. The link between migration and the editing of volumes in the North of theories of the South is central to that editorial process. Such scholars do not stand alone. The "post-" of ethnic and postcolonial studies has seen the rise of different intellectual assemblages of the formerly peripheral (as in Native American studies, Amerindian studies, Black studies, postcolonial studies, not to mention the proliferation of volumes on anthropology from the South, sound studies from the South, popular music studies from the South, and so on, in different languages and from different parts of the world). So the problem posed here means attending not only to the reductive genealogies of the field as noted by Feld (2012a, 2015) and Szendy (2015). It means attending also to the implications of the fact that the South is no longer marginal in the sense of being simply symbolic or an elsewhere that is carefully confined to avant-garde or folkloric metaphorizations, and symbolic inclusions. Rather, the reluctance of a global framework that does not alter dominant accounts generates its own apartheid. This apartheid happens through a politics of confinement, a walling provoked by the conundrums faced by the need to name a separate intellectual space to articulate a presence amid declarations of fluidity and globalization. While these fields are central to challenging the canon, in having to remain peripheral or needing to call attention through

*Ana María Ochoa Gautier*

the very field they name, they speak volumes of the reluctance of disciplinary transformations. This silencing of the South in times of neoliberalization is a form of abandonment (Povinelli 2011), an anesthetic to the Southern economies of the sensorial (Buck-Morss 1992; Ramos 2010).

Loudness (chapter 5), hopping (chapter 1), the liminology of the audible (chapter 4), displaced listening (chapter 10). . . . This afterword opens with place names and closes by calling attention to the stories, people, scenarios, and sound practices that appear throughout the book. One chapter after another in this volume enacts an internal displacement of sound studies simply by attending to the everyday of sound that emerges here. It is in the very storytelling in sound (Feld 2015; Guilbault 2014; see also chapter 10 in this volume) that one reads in each chapter that the crux of the critique emerges. Here is where the experiences in sound multiply; where the lives take over the theory about which this volume speaks. If you read the book from cover to cover, it is overwhelming, uncontainable.

Intellectual cartographies, writes Jesús Martín-Barbero (2005), often involve the artisanal exercise of the chronicler of collective routes. We can think of this as an intellectual cartography that asks us to pay attention to the many possibilities of thinking in sound in and from the South—one that is less a map and more a conglomeration of stories. Much thinking in the South through sound happens not in academic writing but in articulations occurring in different formats and exercises: exchangeable hard drives, blogs, audiovisual compositions, organized action, forms of public gesture, and publications in different languages, among others. There is a parallelism between the different iterations of the South attended to here and the internal displacement of the field these different exercises in thinking and format provoke. This is not about recognition. This is about "tearing down, dismantling, breaking apart"; about a refusal of confinement (Halberstam 2013; Harney and Moten 2013). The invitation of this volume is to disassemble and reassemble, from the South, in the South, and through the South, the privileged montage of sound studies as understood in the North through the practices of thinking the lives, and the nets (Escobar 2008), that are woven in sound and the threads that are used to weave them.

### Notes

1. Thus, for example, the prominent role of the anthropology of music and the anthropology of sound in challenging such taken-for-granted notions as music, sound, and the relation between the human and nonhuman since the early 1980s are hardly ever mentioned as foundational to the field. For recent work from Latin America and

the Caribbean articulated *as* sound studies, see, among others, Campos Fonseca 2014; Estévez Trujillo 2008; Iazzetta 2015.

2. In fact, one could say that the anthropology of sound, by attending to ritual as a central sphere of alteration of what is meant by sound, historically anticipated a later critique of the relation between sound and religion (see, for example, Feld, Menezes Bastos, and Seeger, to name a few).

## References

Buck-Morss, Susan. 1992. "Aesthetics and Anaesthetics: Walter Benjamin's Artwork Essay Reconsidered." *October* 62: 3–41.

Campos Fonseca, Susan. 2014. "Arqueologías sonoras del presente?" *Boletín Música, Casa de las Américas* 37: 27–43.

De Alencar Jacques, Tatyana. 2016. "Rereading *A Musicológica Kamayurá* by Rafael José de Menezes Bastos: 40 Years Beyond an Anthropology without Music and a Musicology without Humans." *TER* 1, no. 1: 240–49.

Escobar Arturo. 2008. *Territories of Difference: Place, Movement, Life, Redes*. Durham, NC: Duke University Press.

Erlmann, Veit. 2004. *Hearing Cultures: Essays on Sound, Listening, and Modernity*. New York: Berg.

Estévez Trujillo, Mayra. 2008. *Estudios sonoros desde la región andina*. Quito, Ecuador: Trama.

Feld, Steven. 2012a. *Jazz Cosmoplitanism in Accra: Five Musical Years in Ghana*. Durham, NC: Duke University Press.

Feld, Steven. 2012b. *Sound and Sentiment: Birds, Weeping, Poetics, and Song in Kaluli Expression*, 3rd ed. Durham, NC: Duke University Press.

Feld, Steven. 2015. "Acoustemology." In *Keywords in Sound*, ed. Matt Sakakeeny and David Novak, 12–21. Durham, NC: Duke University Press.

Feld, Steven, and Donald Brenneis. 2004. "Doing Anthropology in Sound." *American Ethnologist* 31, no. 4: 461–74.

Grosz, Elizabeth. 1999. *Becomings: Explorations in Time, Memory and Futures*. Ithaca, NY: Cornell University Press.

Guilbault, Jocelyne. 2014. *Roy Cape: A Life on the Calypso and Soca Bandstand*. Durham, NC: Duke University Press.

Halberstam, Judith [Jack]. 2011. *The Queer Art of Failure*. Durham, NC: Duke University Press.

Halberstam, Jack. 2013. "The Wild Beyond: With and for the Undercommons." In *The Undercommons: Fugitive Planning and Black Studies*, ed. Stefano Harney and Fred Moten, 2–13. Wivenhoe, UK: Minor Compositions.

Haraway, Donna. 2016. *Staying with the Trouble: Making Kin in the Chthulucene*. Durham, NC: Duke University Press.

Harney, Stefano, and Fred Moten. 2013. *The Undercommons: Fugitive Planning and Black Studies*. Wivenhoe, UK: Minor Compositions.

*Ana María Ochoa Gautier*

Helmreich, Stefan. 2007. "An Anthropologist Under Water: Immersive Soundscapes, Submarine Cyborgs, and Transductive Ethnography." *American Anthropologist* 34, no. 4: 621–41.

Hilmes, Michele. 2005. "Is There a Field Called Sound Culture Studies? And Does It Matter?" *American Quarterly* 57, no. 1: 249–59.

Holbraad, Martin, Morten Axel Pedersen, and Eduardo Viveiros de Castro. 2013. "The Politics of Ontology: Anthropological Positions." Position paper for a roundtable discussion at the Annual Conference of the American Anthropological Association, Chicago, November 21.

Iazzetta, Fernando. 2015. "Estudos do som: Um campo em gestação." *Centro de Pesquisa e Formação*, no. 1: 146–60.

Maniglier, Patrice. 2006. *La vie enigmatique des signes: Saussure et la naissance du strucutralisme*. Paris: Scheer.

Martín-Barbero, Jesús. 2005. *Oficio de cartógrafo: Travesías latinoamericanas de la comunicación en la cultura*. Mexico City: Fondo de Cultura Económica.

Moten, Fred. 2003. *In the Break: The Aesthetics of the Black Radical Tradition*. Minneapolis: University of Minnesota Press.

Ochoa Gautier, Ana María. 2014. *Aurality: Listening and Knowledge in Nineteenth-Century Colombia*. Durham, NC: Duke University Press.

Porcello, Thomas, Louise Meintjes, Ana María Ochoa Gautier, and David. W. Samuels. 2010. "The Reorganization of the Senses." *Annual Review of Anthropology* 39: 51–66.

Povinelli, Elizabeth. 2011. *Economies of Abandonment: Social Belonging and Endurance in Late Liberalism*. Durham, NC: Duke University Press.

Ramos, Julio. 2010. "Descarga acústica." *Papel Máquina* 2, no. 4: 49–77.

Richard, Nelly. 2001. *Pensar en/la postdictadura*. Santiago, Chile: Editorial Cuarto Propio.

Seeger, Anthony. 1987. *Why Suyá Sing: A Musical Anthropology of an Amazonian People*. New York: Cambridge University Press.

Smith, Mark M. 2004. "Introduction: Onward to Audible Pasts." In *Hearing History: A Reader*, ed. Mark Smith, ix–xxii. Athens: University of Georgia Press.

Steingo, Gavin. 2016. *Kwaito's Promise: Music and the Aesthetics of Freedom in South Africa*. Chicago: University of Chicago Press.

Stolze Lima, Tânia. 1999. "The Two and Its Many: Reflections on Perspectivism in a Tupi Cosmology." *Ethnos* 64, no. 1: 107–31.

Strathern, Marilyn. 2000. "Introduction: New Accountabilities" In *Audit Cultures: Anthropological Studies in Accountability, Ethics and the Academy*, ed. Marilyn Strathern, 1–18. London: Routledge.

Strathern, Marilyn. 2004. *Partial Connections*, updated edition. Walnut Creek, CA: Altamira Press.

Strathern, Marilyn. 1988. *The Gender of the Gift: Problems with Women and Problems with Society in Melanesia*. Berkeley: University of California Press.

Sterne, Jonathan. 2003. *The Audible Past: Cultural Origins of Sound Reproduction*. Durham NC: Duke University Press.

Sterne, Jonathan. 2012. "Sonic Imaginations." In *The Sound Studies Reader*, 1–17. Durham, NC: Duke University Press.

Szendy, Peter. 2015. "The Auditory (Re)turn." In *Thresholds of Listening: Sound, Technics, Space*, ed. Sander Van Mass, 18–29. New York: Fordham University Press.

Viveiros de Castro, Eduardo. [2003] 2015. "And." In *The Relative Native: Essays on Indigenous Conceptual Worlds*, 39–54. Chicago: HAU Books.

Viveiros de Castro, Eduardo. 2004. "Perspectival Anthropology and the Method of Controlled Equivocation." *Tipití* 2, no. 1: 3–22.

Viveiros de Castro, Eduardo. 2010. *Metafísicas caníbales: Líneas de antropología postestructural*. Buenos Aires: Katz.

Weheliye, Alexander G. 2014. *Habeas Viscus: Racializing Assemblages, Biopolitics, and Black Feminist Theories of the Human*. Durham, NC: Duke University Press.

*Ana María Ochoa Gautier*

# Contributors

TRIPTA CHANDOLA is an urban ethnographer and Postdoctoral Fellow at the Indian Institute for Human Development in Delhi. In her research experience, spanning more than fifteen years, she has explored different facets of the everyday in the lives of those living on the margins through different tropes and lenses of engagement. She holds a doctorate from Queensland University of Technology, Brisbane, Australia, has published several journal articles and book chapters, and coproduced the BBC radio documentary *Govindpuri Sounds*, based on her doctoral research. She is the author of *Listening into Others: An Exploration in Ethnography* (2019, Institute of Network Cultures, Amsterdam).

MICHELE FRIEDNER is Assistant Professor of Comparative Human Development at the University of Chicago and the author of *Valuing Deaf Worlds in Urban India* (2015, Rutgers University Press).

LOUISE MEINTJES is Associate Professor of Music and Cultural Anthropology at Duke University and the author of *Sound of Africa! Making Music Zulu in a South Africa Studio* (2003) and *Dust of the Zulu: Ngoma Aesthetics after Apartheid* (2017), both published by Duke University Press.

JAIRO MORENO teaches music theory at the University of Pennsylvania. He is the author of *Musical Representations, Subjects, and Objects* (2004, Indiana University Press) and *Syncopated Modernities: Musical Latin Americanisms in the US, 1978–2008* (forthcoming, University of Chicago Press).

ANA MARÍA OCHOA GAUTIER is a Professor in the Department of Music at Columbia University. Her work concerns the global politics of music circulation and histories and anthropologies of aurality in Latin America and the Caribbean.

MICHAEL BIRENBAUM QUINTERO is Assistant Professor of Musicology and Ethnomusicology at Boston University. He writes about the music of black Colombians, cultural politics, violence and trauma, black cosmopolitanism, and vernacular uses of technology. He is the author of *Rites, Rights and Rhythms: A Genealogy of Musical Meaning in Colombia's Black Pacific* (2018, Oxford University Press).

JEFF ROY is a Postdoctoral Fellow with the Centre d'Études de l'Inde et de l'Asie du Sud for "*Projet Autoritas:* Modes d'Autorité et Conduites Esthétiques de l'Asie du Sud à l'Insulinde." He holds a doctorate in ethnomusicology from the University of California, Los Angeles, and has taught in the Department of Communication

Studies at California State University, Northridge. His work focuses on transgender and *hījṛā* music and dance through the lens of documentary filmmaking. Roy has recently published articles in *Ethnomusicology*, *MUSICultures*, and *Transgender Studies Quarterly*.

JESSICA A. SCHWARTZ is Assistant Professor of Musicology at the University of California, Los Angeles. She is the author of *Radiation Sounds: Marshallese Music and Nuclear Silences* (forthcoming, Duke University Press) and is working on a book on listening practices and technologies shaped by the presence of nuclear weaponry in the United States. She is the cofounder and academic advisor of the Marshallese Educational Initiative, an Arkansas-based nonprofit.

SHAYNA SILVERSTEIN is Assistant Professor of Performance Studies at Northwestern University. Her research examines the intersection of sound and movement as modalities of performance, with a focus on contemporary Syrian culture and society.

GAVIN STEINGO is Assistant Professor of Music at Princeton University. He is the author of *Kwaito's Promise: Music and the Aesthetics of Freedom in South Africa* (2016, University of Chicago Press). With Jairo Moreno, he edits the Critical Conjunctures in Music and Sound book series for Oxford University Press.

JIM SYKES is Assistant Professor of Music at the University of Pennsylvania. He is the author of *The Musical Gift: Sonic Generosity in Post-War Sri Lanka* (2018, Oxford University Press) and is working on a book about Hindu sounds, labor, and capitalism in Singapore.

BENJAMIN TAUSIG is Assistant Professor of Ethnomusicology at Stony Brook University. His research focuses on sound, protest, and Asian modernity, and he is the author of *Bangkok Is Ringing: Sound, Protest, and Constraint* (2018, Oxford University Press).

HERVÉ TCHUMKAM is Associate Professor of French and Francophone Postcolonial Studies and associate of the John G. Tower Center for Political Studies at Southern Methodist University. He is the author of several articles and book chapters, coeditor of a volume on postcolonial exile and migrations, and editor of two special issues of scholarly journals devoted to postcolonial Francophone studies and to philosophy and the social sciences. He is the author of *State Power, Stigmatization, and Youth Resistance Cultures in the French Banlieues: Uncanny Citizenship* (2015, Lexington Books).

# Index

abandonment, 16, 135, 139–45, 271

ability, 156, 169, 178–79

Abon, Henson, 99–100

academic production, politics of, 262–63

Achebe, Chinua, 10

acousmatic trace, 214–15

acoustemology, 13, 124, 266

acoustic, 7, 16, 43, 66–67, 69–71; assemblage, 102n3, 269; cleansing, 215–16; cocoon, 43, 53; ecology, 228; epistemology, 124; excess, 15, 19–20, 139, 145–46, 148–49, 269

Adorno, Theodor, 41

*aelōñ kein*, 82

aesthetics, 6, 26, 67, 71, 242, 245–50

affective turn, 54

affordance theory, 159–61, 166, 169–70

Africa, 2–3, 6–7, 14, 16, 39–41, 48, 51, 53–55, 185–94, 196–99

Afro-Colombian social movement, 138

Agamben, Giorgio, 138, 186, 189, 194; the coming community, 186, 198–99

Agawu, Kofi, 8

Ahmed, Sara, 23, 243

*ainikien*, 82, 89

Alessandrin, Patrick, 192

Alesso, Dennis, 93

Algerian War, 69, 194

All Indian Institute for Speech and Hearing, Mysore (AIISH), 162

al-Mashriq, 244

Ambos, Eva, 219

American Academy of Pediatrics, 120

American Board of Commissioners for Foreign Missions (ABCFM), 85, 93

Amerindian studies, 270

Amimoto Ingersoll, Karin, 81, 83

amplification, 16, 64, 70–71, 136, 144, 147, 161, 165

analogism, 111, 121–22, 129

analogue videocassettes, 15

ancestral depth, 79–80, 83

"Andak Bahriya" (song), 251

Android app, 97

*anemkwōj*, 96

animal listening, 218

animism, 122

antenatality, 110, 119–23; and aurality, 110–11, 123, 125–29; and temporality, 120; and vitality, 109

anthropological archive, 265

anthropology, 6, 9, 13, 17, 112, 117, 127, 156, 265–68, 270

apartheid, 52–53, 65, 70, 186, 197, 207

Arab nationalist movements, 245

Argenti-Pillen, Alex, 8, 215–16

Asad, Talal, 210–11

Äsala Perahera, 213

Asia, 6, 14, 25, 86, 140, 157, 187; East, 50; Northeast, 4; South, 3, 21, 179; Southeast, 3, 68

Asociación de Parteras Unidas del Pacífico (ASOPARUPA), 113, 124

Astley, Tom, 52

astrology, 22–23, 205–7, 216, 221

*'ataba*, 245

Atomic Energy Commission, 87

atomization, 41

Attali, Jacques, 5

audibility, 12, 19, 21–22, 185, 192, 194, 242, 257, 269

audile practices, 109

audile techniques, 114, 248. *See also* technique

audiovisual litany, 2–3, 7, 53, 209, 211, 213, 221, 263

auditory technology: cochlear implants, 159, 161–62; hearing aids, 47, 49, 159, 161–63

aural fantasies, 256

aurality, 21, 83, 110–11, 120–21, 123–29, 173, 209, 243, 256

auscultation, 109, 111–12, 114, 125, 129

authenticity, 96, 176

automobile listening, 43, 51

autonomy, 42, 255

*awwal*, 245, 249–50, 253–54

Ayurvedic medicine, 213

*badhai*, 177–78

Bakhtin, Mikhail, 68

*bali tovil*, 212

Balibar, Étienne, 199

ballet, 241, 252–54

Bandaranaike, S. W. R. D., 217

Bandung Conference (1955), 25

bandwidth, 45, 52, 64, 67

Bangalore, India, 156, 160–62, 261

Bangkok, Thailand, 156, 164, 166–67, 169, 261

*banlieues*, 4, 22, 185–99

*Banlieue 13 — Ultimatum* (2009 film), 192, 195

*bao in lal*, 92

bare life, 40, 138

Barkat, Mohamed Sidi, 196

Bataille, Georges, 148, 150

Bate, Bernard, 9

*Battle of Algiers, The* (1966 film), 69

Baucom, Ian, 10, 27

Becker, Judith, 9

Beer, David, 49–50

Belmessous, Hacène, 195

Benna, Zyed, 191

Bera Poya Hevisi, 217–18

Beravā caste, 212–13, 215, 217–18, 220

Bessire, Lucas, 10

Bhattacharya, Sayan, 180

Bickford, Tyler, 50

Bijsterveld, Karin, 43–44, 46–48

Bikini Atoll, 87, 92–93

biology, 20, 87, 109–10; as basis of life and death, 117, 119, 123, 125, 127–28

biomedicine, 110–11, 113–15, 120, 124–25

biopolitics, 16, 41, 46, 49, 52, 101, 117, 125, 142–43

biopower, 47, 52–53, 136, 140–41, 143

birds, 4, 9, 16, 78, 81, 89, 91–92, 94, 100–101

Birla, Ritu, 212

black Atlantic, the, 71

Black France, 198

black studies, 16, 71, 269–70

Blavatsky, Helen, 220

Bluetooth, 48–52

body hearing aid (Siemens), 161

body-voice, 64–65, 70–71

Bonenfant, Yvon, 177

Bongela, Milisuthando, 13

Boredoms (band), 203–4, 221

Born, Georgina, 5

Bosavi rainforest, Papua New Guinea, 228

*botella curada*, 112

Boubeker, Ahmed, 192–93, 198

Bourdieu, Pierre, 252

Brathwaite, Kamau, 9

Bronfman, Alejandra, 11

Buddhism, 22, 218–21

Budh Bazaar, 235

Buenaventura, Colombia, 112, 114, 135, 138–42, 145, 149, 261; midwives in, 120, 125

Bull, Michael, 6, 43–44, 49, 51

canoes, 4, 16, 77–78, 80–84, 86–91, 93–94, 98, 100–101
Capitaine Danrit (Émile Driant), 187, 198
capoeira, 241
captioning, 159
Carpenter, Dale, 91
Carpenter, Edmund, 3
cartography, 4, 11, 24–25, 78, 80–81, 85, 111, 127, 129, 261, 266, 269–71
celestial jukebox, 44–45, 52
Césaire, Aimé, 188–9
Chakrabarty, Dipesh, 127
Chandola, Tripta, 19, 23, 178
Chatterjee, Partha, 142–43, 212
chigualo, 117–18
Chion, Michel, 114–16
Chirac, Jacques, 193
Chong-Gum, Carmen, 91, 94–95
Chow, Rey, 50–51
Cimini, Amy, 53
circulation, 11–12, 16, 41, 51–52, 101
citation, politics of, 263, 270
classical music and dance, 174–78, 251
climate change, 88, 99, 101. *See also* global warming
cloud (technology), 45, 48
Colombian civil conflict, 141
Colombian Ministry of Culture, Division of Patrimony, 113
Colombian Ministry of Health, 113
colonial Ceylon, 212, 220
colonial continuum, 186
colonialism, 3–4, 22, 24, 65, 78, 189, 213
colonial maps, 81
*Colonizer and the Colonized, The* (Memmi), 194
*communautarisme*, 188
Compact of Free Association (COFA), 87–88, 100
Conference of Berlin (1884–85), 86
conjunctural sound studies, 242
constitutive technicity, 11, 12–15

Corbin, Alain, 137, 209
*corps d'exception*, 196
Council of Iroij, 88, 101
Coutant, Isabelle, 196
Crawley, Ashon, 71
Csikszentmihalyi, Mihaly, 242
culturalism, 21
cultural relativism, 124
cultural revitalization, 89
cunning of recognition, 192
cutting houses (*casas de pique*), 147, 149–50, 153n14

*dabbik*, 249, 252–54
*dabha*, 236–37
*dabke*, 23–24, 241–56
*dal'ouna*, 245
Damascus, Syria, 241, 253, 261
Dassanayake, D. M., 214, 217
deafness, 20, 156–60, 164, 166, 169, 190, 228, 256, 264, 267
deaf sociality, 161, 164
deaf studies, 159
decolonization, 4–5, 12, 24, 100, 129, 198, 264
deconstruction, 3, 25n6, 125, 262
Delhi, India, 23, 229, 235
DeLoughrey, Elizabeth, 81
Department for International Development, India, 229
Derrida, Jacques, 3; *l'arrivant*, 118–19; life/death, 118, 125, 128
Descola, Philippe, 110–11, 121–22, 127, 129
de Silva, Premakumara, 218
*desouvrée*, 146
*devales*, 205
*deva tovil*, 212, 218
Dhamma, 205, 214–15, 218
*dharmadesana*, 213
Dharmapala, Anagarika, 213
dialogic editing, 71
dialogism, 71

diaspora, 81, 100; African, 113, 198; indigenous, 78, 80; Marshallese, 16, 77–78, 81, 89, 93, 101; Pacific, 100
Dik, Ali El-, 246
disability, 20, 156–59, 167, 169–70
Disabled People's Association, Bangalore (DPA), 161
disciplines, 5–6, 11–12, 27n26, 41, 53, 126, 262–65
disciplinization, 261–62
*Discourse on Colonialism* (Césaire), 188
disfiguration, 24, 264–65, 267
disidentification, 176
displaced people, 138
*dispositif* of recognition, 128
divine powers, 110, 120
Dreyfus affair, 187, 199n3
*droit de cité*, 199
Du Bois, W. E. B., 71
Dubreuil, Laurent, 190
du Gay, Paul, 42
*dukkha*, 220
Dutta, Aniruddha, 174, 180

earbuds, 50
earplugs, 47, 49
ecomusicology, 6, 132n24, 223n13
efficacious ritual, 213, 218
Eisenberg, Andrew, 10
Ejit Island, 93
*ejjeḷọk wōnen*, 91
Elbōn, Antari, 98–99
Ellison, Ralph, 192
embodied experience, 242, 256
embodied knowledge, 242–43, 254
embodied practice, 70, 249, 252, 256
embodied sensibility, 249
empathic kinesthetic perception, 252–54
endurance, 117, 148
Enewetak Atoll, 86–87
Enlightenment, 55
Ensoniment, 55

Enthoven, Raphaël, 191
Episos, Alfred, 84
equivocation, 24, 264–69
Erlmann, Veit, 3, 21, 55, 209
ethical affordances, 160
ethnographic encounter, 23–24, 243, 249, 251, 254, 256–57
ethnography, 4–5, 24, 71, 111, 244, 264–65
ethnomusicology, 6–7, 96, 266
ethno-nationalism, postcolonial, 208, 219
ethno-racialization, 188
everyday politics, 138–39, 143, 146
excess, 15, 19–20, 53, 139, 145–46, 148–49, 269
extensive differences, 123, 125, 127
Eye, Yamataka, 203, 207

factory sound, 46–47, 147
Fanon, Frantz, 26, 193
Fassin, Didier, 188
Fassin, Éric, 188
Feld, Steven, 13, 71, 124, 228, 265, 270
feminist theory, 269
Ferguson, James, 48–49, 54–55
fiduciary logic, 53–54
file sharing, 45, 52
Fiore, Quentin, 53
Fisher, Daniel, 10, 73n7
flow, 14, 48, 52, 55, 97, 242, 249–50, 252
folk dance, 245
foreigners: *étrangers*, 186–87; postcolonial, 187
Foster, Don, 69–70
Foucault, Michel, 47, 52–53, 125, 136
Françafrique, 189, 193, 196
Friedner, Michele, 19–20, 156, 158–60, 267
Fuerzas Armadas Revolucionarias de Colombia (Revolutionary Armed Forces of Colombia), 150
Fundación Activos Culturales del Afro (ACUA), 113

funerary practices, 7, 117, 119
fungibility, 14–15

Gallope, Michael, 11, 15
*ganachandas*, 217
Garcia, Luis-Manuel, 248
gender, 7, 9, 12, 16, 20, 23, 146, 173–77,
    230–31, 234; in *dabke*, 243–45, 247;
    midwifery and, 113; in *ngoma*,
    63–71
*gharanas*, 178–79
gift, 109, 120, 166–69, 207, 217
Gilbert, Thomas, 85
Gilbert Islands, 85
Gilroy, Paul, 71
global cities, 6, 22, 135, 211
global convergence narrative, 46, 55
global modernity, 4–5, 7, 12, 52, 135, 141,
    211
global North, 14–15, 46, 51, 55, 80, 112, 137,
    156, 167–68, 170, 175
global South, 1–16, 21–23, 46, 52, 55,
    63–64, 71–72, 138–39, 156–57, 185–86,
    208
global warming, 78, 88. *See also* climate
    change
gold mines, 16, 52, 148
Gondola, Didier, 198
Goodman, Steve, 17
Gopi (singer), 176–77
Gopinath, Sumanth, 46
governmentality, 124–25, 136, 141–43
Govindpuri (lower-middle-class neigh-
    borhood), 234–35
Govindpuri slums, 23, 228–32, 235,
    238
GPS (Global Positioning Systems), 77, 83
Graeber, David, 157, 159–60, 170
grammatology, 267
Guéant, Claude, 189, 193
*guru-chela*, 178
gynecology, 114

Habib, Wafik, 246, 251
*ḥaflāt*, 241, 247–48, 255–56
Hägglund, Martin, 118
Harris, Elizabeth, 219–20
Hartblay, Cassandra, 158
Hau'ofa, Epeli, 83
Haupt, Adam, 51
hearing loss, 16, 46–49, 52
Heller, Michael, 147
Henos, Franklein, 98
heteronormativity, 179–80
hījṛā, 20, 173–74, 176, 178–80
hīj-vocality, 176–80
Hilmes, Michele, 5–6
hip-hop, French, 185, 190
Hirschkind, Charles, 9, 208–9, 211, 255
historiography, 21, 264
HIV epidemic, 65
Holbraad, Martin, 24, 126, 264
Holmquist, Lars, 44–45
Holt, John, 221
home sound systems, 19. *See also* sound:
    systems
*homo sacer*, 194
Hosokawa, Shūhei, 42
*hu dteung*, 165
Hurston, Zora Neale, 71
Husserl, Edmund, 147

Ihde, Don, 147
improvisation, 72, 245
inaudibility, 22, 185, 194
inclusive exclusion, 189
India, 7, 14, 20–21, 156–64, 173–76, 189, 219
Indian Ocean tsunami (2004), 216
Indian Sign Language (ISL), 158
indigeneity, 21, 78–81, 83–89, 96, 121, 210,
    218–20; indigenous navigation, 87, 89;
    and relationality, 80; and sovereignty,
    81
Ingold, Timothy, 2–3, 243
intellectual labor, 263–65, 268–69

intensive differences, 123, 125
intercorporeality, 72
International Disability Alliance, 158
International Monetary Fund, 3
International Phonetic Alphabet, 94
intersectionality, 110
intersubjectivity, 137, 252, 257
intervocality, 71–72
invisibility, 22, 78, 81, 142–43, 146–47, 180, 185, 192, 195
*Invisible Man* (novel), 192
*ippān doon*, 91
irrationality, 2, 136, 139, 142
iTunes, 40
*izzat*, 20, 173–74, 178–79

Jackson, John, 174
Jaluit Atoll, 86, 90
*jambo*, 92
James, Kery (rap artist), 191
Janatha Vimukthi Peramuna (JVP), 215
*jebta*, 94–96
*jhuggi*, 229, 234–35, 239n3
Jiménez, Orián, 10
*jitdaṃ kapeel*, 89, 93, 96
Johannesburg, South Africa, 44, 51, 61
Jordan (country), 244
Judeo-Christian theology, 2, 208

Kabua (Marshallese principal chief), 86
Kaluli acoustemology, 13
Kandyan Dance, 213, 219
Kano, Nigeria, 15
Karam, Fares, 246
Kassabian, Anahid, 14, 45–46, 53
Keane, Webb, 160
Kedi, Kenneth, 88
Kelen, Alson, 93
Khan, Faris, 179
Kheshti, Roshanak, 96, 256
*Khwaja Sira*, 180
Kili Island, 93

kinesthetics, 23, 243, 252–55
Kirkley, Christopher, 51
KMRW 98.9 FM (Springdale, Arkansas), 97, 99
Kohn, Eduardo, 127, 129
Kohomba Kankariya (healing ritual), 213, 219
Koolhaas, Rem, 55
Koovagam festival, Tamil Nadu, 176–77
*koṭhīs*, 177
"Kotta Paakkum" (song), 176–77
Kracauer, Siegfried, 41
Krell, Elias, 175
Kuipers, Joel, 67, 69–70
Kunreuther, Laura, 11
Kūrijmōj season, 94
Kwajalein Atoll, 86, 90, 98
KwaZulu-Natal, 61, 63, 261

LaBelle, Brandon, 43
labor migration, 61, 65, 187
language(s), 1, 19, 54, 71, 101, 110, 120, 126; engagement with scholarship across multiple, 4–6, 10, 270–71; insufficiency of, 149; Marshallese, 82–85, 89, 94–95; rioting as, 190, 194, 199; sign, 156–61, 163; spoken, 19, 94
Lapeyronnie, Didier, 186, 189, 193–95, 198
Larkin, Brian, 10, 15, 243
lateral agency, 150
Lattakia, Syria, 249, 251
League of Nations' Mandate System, 86
Lebanon, 244
Lee, Tong Soon, 10
leftist guerrillas, 112, 142
Leppert, Richard, 148
lettered city, 218, 220
Lévi-Strauss, Claude, 3, 170n1
liberal: citizenship, 135; language, 101; modernity, 137; multiculturalism, 125

liberalism: economic, 88; political, 21, 137–39, 208, 212–13

Liberation Tigers of Tamil Eelam, 214

Limbe, Cameroon, 39, 44, 48

liminology, 18–19, 135, 139, 271

limits of sound, 5, 17, 19–20, 66

*L'invasion noire* (novel), 187, 198

listening: into others, 179, 237; practices, 19, 50, 114, 238, 243, 248, 256; subject, 6, 12, 15, 41, 114

Livermon, Xavier, 16, 51

Loeak, Daisy, 92

Losonczy, Anne-Marie, 120

Losurdo, Domenico, 137

MacDonald, Barrie, 85

Mackenzie, Adrian, 267

Madariaga, Patricia, 143

Madhumitha (singer), 177

*madhya saptak*, 178

Madrid, Alejandro L., 174

magic, 2, 84, 212, 215–16

Mahmood, Saba, 210–11

Majuro Atoll, 86, 90, 93, 97–99

Manabe, Noriko, 137

managed population, 143

*mantin Majōl*, 89, 91

*mantras*, 212, 215–16

March for Equality and Against Racism, France (1983), 185

Marsden, Magnus, 9

*Marshallese-English Dictionary*, 94

Marshallese navigational chants, 84, 91, 93, 96. See also *roros*

Marshall Islands, 16, 78–101

Martín-Barbero, Jesús, 10, 271

masculinity, 48, 61, 66, 91, 109, 212, 241, 254–55

*mawwal*, 245, 250

Mbembe, Achille, 8, 52–55, 136, 191, 196

McLuhan, Marshall, 3, 53

medicinal plants, 112, 121

Mediterranean South, 1

Meintjes, Louise, 16–17, 61–72, 267

Memmi, Albert, 194

midwives, 4, 19, 109–29

migrants, 22, 61, 65, 185–88, 270. *See also* labor migration

Mills, Mara, 47

Milne, Simon, 91

mimesis, 176, 178, 252

miners, 49, 52–53, 148

Minh-ha, Trinh T., 9

miniaturization, 42, 47, 50

missionaries: accounts of, 69; in the Marshall Islands, 82, 84–86, 94; in Sri Lanka, 22, 208, 219–21

*mittasil*, 250, 254

*mizān*, 253–54

mobile: listening devices, 15, 41–43, 47; music, 46, 49, 56n7, 80, 82, 97; phones, 44, 51, 231; privatization, 41

mobility, 14–15, 41–42, 44, 71, 81, 86–87, 96, 101, 231, 234; economic, 156, 168–69

Modi, Narendra, 158

Moisa, Lebona, 10

Molano, María Elvira, 110

Morel, Pierre, 195

Moreno, Jairo, 18–19, 53, 109–29, 146, 267

Morris, Jeremy, 45, 48

Morris, Rosalind, 49

Moten, Fred, 71, 269

MP3: format, 40, 48, 52; player, 45, 50

Msinga, South Africa, 61, 65

Muller, Larry, 97

multiculturalism, 18, 111, 125, 199

multinaturalism, 18

Muñoz, José Esteban, 176, 179

musicology, 7, 206

*muwashshaḥ*, 245

Muzak, 45

Mwekto, Willie, 83

Mxit, 14, 51

naming, 85, 92, 269
"Namo Namo Matha" (Sri Lanka national anthem), 217
Nancy, Jean-Luc, 116, 144, 146
natal charts, 4, 205
National Telecommunications Authority, Marshall Islands (NTA), 89–90
National Thai Sign Language, 157
native, 2, 21, 187, 193, 221, 270
Native American studies, 270
nativism, 14
naturalism, 110–11, 124, 126, 129
Navarro Valencia, Martha Cecilia, 120
navigation, 4, 12, 78, 80–81, 83–89, 93, 98, 100, 266
Navjeevan Camp, 230
necropolitics, 138, 141–43, 146, 150
neoliberalism, 3, 101, 135, 139, 228, 246, 262–63, 271
network archaeology, 100
New Island: The Marshallese in Arkansas, A (film), 91–92
ngoma, 61–67
Nitijela, 88
noise, 6, 44, 46–47, 63, 82, 137–38, 145, 165, 167–68, 210, 212, 264
nonhuman animals, 14, 17, 122, 207, 218–19
not-self, 217
Novak, David, 26n12, 28n32
Nuclear Claims Tribunal, 88
nuclear colonialism, 78
nuclear weapons tests, 87, 101
Nzewi, Meki, 9

O'Callaghan, Casey, 116
oceanic literacy, 83
Ochoa Gautier, Ana María, 9, 11, 19, 209, 212, 218, 261–71
ocularcentrism, 208, 212, 268
Ong, Walter, 3, 209
online music store, 45

Operation Castle Bravo (1954), 87
oral genres: Marshallese, 86, 89
orality, 173, 209, 220
Ott, Katherine, 158
Ozarks, 99

Pacific Afro-Colombia, 19, 109–29
padas, 218
Palestine, 3, 244
Pandora (online radio), 40
para-human, 111, 130n4
paramilitaries (Colombia), 112, 142–43, 146–47, 152n9
Paribatra, Sukhumbhand, 164
parlure, 190
partial connections, 268–69
Peake, Bryce, 243
Pedersen, Morten Axel, 126
perahera, 212–13
performance ethnography, 24, 244
performance theory, 54
personhood, 135
Persons with Disabilities (Equal Opportunities, Protection of Rights, and Full Participation) Act (India), 158
Persons with Disabilities' Quality of Life Promotion Act (Thailand), 157
perspectivism, 18, 111, 122
Petty, Karis, 161
phenomenology, 54, 110, 112, 114, 117, 243
philosophy of music, 7
phonograph, 82, 96
Piedras Cantan (neighborhood), 145
Pietz, William, 10
Pinard horn, 112, 114–15
pirit, 214–15
playlist, 23, 40, 45
polyphony, 64, 68, 71
Portela Guarín, Hugo, 110
postcolonial critique, 262
posthumanism, 18, 28n35
postnatality, 110–11, 123, 128

potlatch, 148
Povinelli, Elizabeth, 18, 96, 117, 192
practical metaphysics, 111, 116, 119, 126
practices of loudness, 135, 138, 150
Prasad, Pavithra, 174
primitive rebellion, 186, 198
private property, 43, 138, 144
private space, 7, 21, 52, 210, 221
privatization, 41, 44, 112, 140, 164
proprioception, 23, 242–43, 252–56
Protestant secularism, 210, 213
provincialization, 127
proximate hearing/remote deafness,
    166–67
public space, 21, 44, 210, 212, 247

quasi-divinity, 117, 119
quasi-event, 117–20, 123
quasi-life, 19, 111–12, 117, 120, 122–23, 127
queerness, 2, 176–77, 179
*Queer Phenomenology* (Ahmed), 23, 243
queer theory, 266, 269
queer vocality, 20, 177
"Qu'est-ce qu'on attend?" (song), 190
Quiñones, Liceth, 113
Quiñones, Rosmilda, 113

racial capitalism, 24–25
racial sincerity, 174
Rajapakse, Basil, 214
Rajapakse, Mahinda, 216
Rama, Ángel, 10
Ramos, Julio, 10
Ratana Sutta, 215
rationality, 2, 128
recognition, politics of, 125, 263
Reddy, Gayatri, 20, 174
reduced listening, 114
refugees, 23, 141, 247
Rehabilitation Council of India, 164
Rehabilitation of Disabled Persons Act
    (Thailand), 157

respectability, 175
Restrepo, Eduardo, 117
Riddle, Charles, 10
Rights of Persons with Disabilities Act
    2016 (India), 158
ringtones, 28n30, 45
Rivera-Servera, Ramón, 174
Roberts, Michael, 10, 212
Rocard, Michel, 193
Rojas, Janet, 130n5
Roman property law, 85
Rongelap Atoll, 87
*roros*, 84, 91, 93, 96
Rousseau, Jean-Jacques, 1–3
Rowell, Lewis, 21
Roy, Raina, 174
*rozana*, 245
rule of law, 138–39

*sabda pujāva*, 217
sacred sounds, 207–8, 210–12
Sakakeeny, Matt, 26n12
*salṭana*, 242, 248–50, 253–55
*samādjaya*, 215
Samarakoon, Ananda, 217
*sammī'a*, 242, 245, 251
Sandton, South Africa, 39
Sangam Vihar, 235
Sarkozy, Nicolas, 198
Sartori, Andrew, 21, 208
satellites, 83, 90, 246
Schafer, R. Murray, 3, 5, 223n13
Schmidt, Leigh Eric, 209
Schwartz, Hillel, 157
Schwartz, Jessica, 16–17
sciences, the, 2, 6–7, 12, 109, 126, 162, 213,
    215
secondary genre, 68, 71
secularism and secularization, 7, 209,
    210–11; and modernity, 2, 220
self, the, 23, 25, 119, 126, 147–48, 175–77,
    207, 211, 217, 221, 229, 242

sensorial emplacement, 161
sensory disorientation, 23–24, 244, 251, 254–56
Sentilraj, Deepa, 162
settler colonies, 3
77 Boadrum, 203–7, 215
sfx Car Audio (garage), 164–65, 167–68
Shamanism, 122
Shannon, Jonathan, 242
shawm, 217, 245
Shelemay, Kay Kaufman, 9
Sheriff, Robin, 149
signifyin(g), 63
sign language, 156–63
Silveira, Lucas, 175, 177
Silverstein, Paul, 196–97
Silverstein, Shayna, 23–24
Simmel, Georg, 41
Sinhala Buddhism, 4, 22, 204, 207–8, 212–21
"Sinhala Only" law (1956), 216
Sirisena, Maithripala, 219
Skafish, Peter, 122
skandhas, 217
Sklar, Deidre, 243, 252
Smith McKoy, Sheila, 69
social media, 14, 246
sociology of science, 6
sonic: accumulation, 16; causation, 54; difference, 4, 208; efficacy, 18, 207, 214, 217, 221; envelope, 53; exchange, 22, 39, 48; history, 12, 21; immersion, 136; protection, 4, 207, 213, 218; technicity, 16
sound: -in-itself, 17; medical approaches to, 7; reproduction, 6–7, 12, 16, 81, 84, 96, 98, 100–101, 211; salvaged, 159, 164; spectrality of, 19, 263; spoiled, 160; systems, 39, 136, 145, 148, 151, 165–67, 247–48, 255; as vibration, 17, 54, 80, 115, 215, 248, 256
Southern Hemisphere, 3

sovereignty, 19, 80–82, 123, 135, 138, 142, 150, 219
Soweto, South Africa, 39–40, 44, 48–51, 54, 261
speciation, 121
speech acts, 7, 9
Spotify, 40
Springdale, Arkansas, 81, 91–97
Sri Lankan Civil War (1983–2009), 213, 217
Stanyek, Jason, 46, 53
Starosielski, Nicole, 83
state of exception, 139
Steingo, Gavin, 1–25, 39–55, 64, 68, 208, 267
Sterne, Jonathan, 2–3, 7, 53, 55, 209, 211, 242, 256, 263
Stewart, James, 218
Steyn, Melissa, 69–70
Stiegler, Bernard, 13
Stoller, Paul, 9
structural adjustment, 3, 55
subjectivity, 24, 110, 144, 147, 173, 175, 244
Sugarman, Jane, 175
superfluity, 16, 53
Suprême ntm (French hip-hop group), 190
Sykes, Jim, 1–25, 64, 137, 203–21, 267
Syria, 241–55; conflict in, 246; nonviolent protests of 2011–12, 246
Szendy, Peter, 3, 270

ṭabl, 245, 248–50
tactility, 23, 243, 256
Tagore, Rabindranath, 217
Tambiah, Stanley, 9
tānni, 245, 249, 253–54
ṭarab, 242, 245, 248
Tausig, Benjamin, 18–20, 156–69, 267
Taylor, Charles, 210
techne, 64
technical prostheses, 13

technicity, 11–12, 16

technics, 113

technique, 11, 63–69, 113, 169, 179, 270; audile, 114, 248–49; body, 24, 252, 255; performance, 241–44; virtuosic, 173–76; vocal, 16, 173–76

technological failure, 14–15

technology problematic, 5, 12–17

téléchargeurs, 40, 48

temporality, 20, 116, 118, 120, 128, 150–51

terra nullius, 80

Tetreault, Chantal, 196–97

Thailand, 20, 156–68; military coup of 2014, 157

"The Canoe Is the People," 77, 80

Theosophical Society, 220

Theravada Buddhism, 22, 207

third gender, 173, 179

timbre, 2, 178

Tomlinson, Gary, 10, 13

tonic sound, 114

townships, South Africa, 16, 39, 51, 186

trance, 7

transduction, 11, 21, 81, 248, 267–68

transformation, 24, 264, 267–71

transgender, 20, 173–80, 180n1; vocality (also trans-vocality), 175, 177

transgender-hījra. See trans-hījra

trans-hījṛa, 20, 173, 176–79

translation, 5, 10, 69, 94, 122, 126

transmission, 16, 83, 267

Traoré, Bouna, 191

tree spirits, 207

Triple Gem (Buddhism), 214

Tsing, Anna, 18

tympanic membrane, 115

Tyson poultry plants, 92, 95

ubiquity of music, 14, 41, 44–55

ululation, 4, 16, 61–72; Levantine, 70; Moroccan, 68, 70; racial, 69; white, 69; Zulu, 4, 61–72

United Arab Republic (1958–61), 245

United Nations Convention on the Rights of Persons with Disabilities (UNCRPD), 157–58

United Nations Decade of Disabled Persons, 157

United Nations Educational, Scientific and Cultural Organization (UNESCO), 80

United Nations Trust Territory of the Pacific Islands (TTPI), 87, 95

univeralism, 44, 157–58, 170, 192, 242

unspeakability of violence, 20, 149

urban riots, 22, 186, 188, 190–91, 193, 198

urban space, 6, 136, 142, 146, 151, 178

USB devices, 40, 52

Väddas, 218

value, 2, 21, 67, 91, 99, 148, 156–57, 159–70, 228, 267

vastu, 215

Vergès, Françoise, 4, 24–25

viche, 112

video phones, 159

Virilio, Paul, 42

virtuosity, 20, 66, 173–80, 266

visibility, 180, 185, 192, 199

visible minorities, 22, 186, 192, 195

Viveiros de Castro, Eduardo, 18, 24, 121–23, 126, 264

vocality, 71–72, 173–80

vulnerability, 39–40, 53, 66, 121

Waan Aelōñ in Majōl (Canoes of the Marshall Islands; WAM), 93

Walkman, 42, 47, 49–51

War on Terror, 137

Wasantha, Rohana, 204–7, 215

waterfalls, 13

wave piloting, 86, 93

Wayamba provincial elections (1998), 214

Weheliye, Alexander, 71
Weidman, Amanda, 9, 173–74
Weinberg, Gil, 45
Weiner, Isaac, 209–10
Western thought: epistemologies, 81–82, 127; exceptionalism, 21; modernity, 2, 21, 208–9, 212; naturalism, 124; reason, 2; transcendental subject of knowledge, 127
whaling vessels, 85
WhatsApp, 14, 235
White, Graham, 209
White, Shane, 209
Williams, Raymond, 41–42

World Bank, 3
World Federation of the Deaf (WFD), 158–59
Wotje Atoll, 90

*yak tovil*, 212–13, 216
*yakku*, 207, 212, 216
*yantra*, 215

Zaffiro, Jim, 10
Zambrano Cuero y Caicedo, Sixta Tulia, 112–15, 119, 124, 126–29
*zones de non droit*, 186
zoopolitics, 218–19

www.ingramcontent.com/pod-product-compliance
Lightning Source LLC
Chambersburg PA
CBHW050338270326
41926CB00016B/3505